Microwave Technologies - Recent Advances and New Trends and Applications

Edited by Hüseyin Şerif Savci

Published in London, United Kingdom

Microwave Technologies - Recent Advances and New Trends and Applications
http://dx.doi.org/10.5772/intechopen.110921
Edited by Hüseyin Şerif Savci

Contributors
Alexander Nechaev, Angie Fasoula, Dalia El Khaled, Francisco Portillo, Grigory Dunaevskiy, Igor Dorofeev, Jose Antonio Gazquez, Lanlan Liao, Luca Olmi, Luc Duchesne, Manuel Fernadez-Ros, Marinus Kunst, Mengjie Qin, Nelly Lara, Nuria Novas Castellano, Pengzhan Liu, Petros Arvanitis, Polina Smygalina, Reinhard Schwarz, Rosa Maria Garcia, Veysel Demir, Xiaozhuang Hu, Xin Qiu, Zhiqiang Li, Zhongmao Li

First published in London, United Kingdom, 2024 by IntechOpen
IntechOpen is the global imprint of INTECHOPEN LIMITED, registered in England and Wales, registration number: 11086078, 167-169 Great Portland Street, London, W1W 5PF, United Kingdom

British Library Cataloguing-in-Publication Data
A catalogue record for this book is available from the British Library

Additional hard and PDF copies can be obtained from orders@intechopen.com

Microwave Technologies - Recent Advances and New Trends and Applications
Edited by Hüseyin Şerif Savci
p. cm.
Print ISBN 978-1-83769-694-9
Online ISBN 978-1-83769-693-2
eBook (PDF) ISBN 978-1-83769-695-6

For EU product safety concerns:
IN TECH d.o.o., Prolaz Marije Krucifikse Kozulić 3, 51000 Rijeka, Croatia,
info@intechopen.com or visit our website at intechopen.com.

Meet the editor

Hüseyin Şerif Savcı received his BS degree in electronics and communication engineering from Yıldız Technical University, İstanbul, Turkey, in 2001, and his MS and Ph.D. degrees in electrical engineering from Syracuse University, Syracuse, New York, USA, in 2005 and 2008, respectively. His dissertation on "Low Power CMOS Receiver for Medical Implant Devices" was a recipient of the Best Thesis Award. From 2008 to 2013, he was with Skyworks Solutions Inc., Cedar Rapids, Iowa, USA, as a senior RFIC design engineer. Between 2013 and 2020, he worked with Hittite Microwave Corporation in Chelmsford, Massachusetts, and Analog Devices Inc. in İstanbul as a principal design engineer. Over the years, he designed and released many RFIC and MMIC products on CMOS, SiGe, SOI, GaN, and GaAs technologies. He is currently an Associate Professor in the Department of Electrical and Electronics Engineering at Istanbul Medipol University, İstanbul. He established the RFMicroSense Research Group, where he conducts research on the design and modeling of RF and microwave integrated circuits, devices, systems, and antennas. He is also an associate editor for the Applied Computational Electromagnetics Society.

Contents

Preface

Microwave engineering, which has had a well-established theory for tens of decades, is one of the leading multidisciplinary engineering fields used in today's technology developments. This book is intended to present some of the recent advances and new applications in diverse uses of microwave technology, from medical applications to radio astronomy, from the food industry to material characterization.

The book consists of eight chapters. Within each chapter, theories coupled with practical examples are reviewed and emerging conclusions are drawn.

In the first chapter, author Nelly Lara introduces the basic principles available in microwave ovens based on the dielectric heating modes. The chapter covers recent advances and challenges in uniform heating using movement or rotating devices. Finally, the pros and cons of applying microwave energy in toasting operations are discussed.

In the second chapter, authors Grigory Dunaevskiy, Polina Smygalina, Igor Dorofeev, and Alexander Nechaev discuss how limb injury due to cold remains a serious and widespread condition both in countries with a cold climate and in regions located close to the equator but having high-altitude territories. To solve this problem, the chapter proposes the use of microwave radiation at an early stage, which penetrates deep into the cooled volume and, accordingly, can accelerate the activation of internal vessels. The chapter gives technical details such as the selection of the appropriate frequency and power of radiation, as well as the creation of a microwave chamber in which it is possible to ensure sufficiently uniform heating of the entire volume of the frostbitten limb. At the end of the chapter, the authors present the results of modeling the distributions of electromagnetic and thermal fields in the volume of the heated limb at microwave frequencies that allow medical applications.

In the third chapter, authors Angie Fasoula, Petros Arvanitis, and Luc Duchesne explain microwave imaging technology for the diagnosis of breast cancer. The chapter presents the methodology employed for the quantitative assessment of the reliability of the Wavelia#2 microwave breast imaging (MWBI) system based on experimental data. The analysis includes an assembly of the most representative findings from the series of the MWBI experimental tests performed on breast phantoms after the installation of the Wavelia#2 investigational device at Galway University Hospital (GUH) Symptomatic Breast Unit for a phase-II pilot clinical investigation. The authors explain that the notion of MWBI scan and reconstructed image repeatability assessment has never been addressed before in the MWBI state of the art.

In the fourth chapter, authors Nuria Novas Castellano, Dalia El Khaled, Rosa Maria Garcia, Francisco Portillo, Manuel Fernadez-Ros, and Jose Antonio Gazquez introduce dielectric spectroscopy (DS), a nondestructive experimental tool to investigate the interaction between a material and an applied electric field by utilizing the spectral range from 10 Hz to 1 THz. The chapter focuses on its use in evaluating the properties of living tissues, especially in the horticultural field. Characterizing the dielectric properties of fresh food enables the evaluation of food quality and freshness from production until consumption. The chapter reviews current research on the dielectric characteristics of fresh foods and the application of these techniques in assessing food quality.

In the fifth chapter, authors Marinus Kunst and Reinhard Schwarz show that the change in the microwave reflection of a semiconductor under illumination could be used to determine photoconductivity. The chapter discusses the reliability of the method experimentally, both on samples with known parameters such as silicon and on samples for which the measurements of systems are difficult to access with conventional photoconductivity tools, such as an amorphous semiconductor, a-Si:H, and a semiconducting powder, TiO_2.

In the sixth chapter, author Luca Olmi explains the growth of radio telescope technology and instrumentation. The author mentions the slow adoption of new antenna technologies, especially for large radio telescopes. The chapter discusses the strengths and weaknesses of specific new technologies, particularly reflect arrays, which have been developed mainly for remote sensing and satellite communications. The author presents that reflect arrays might lead to a great leap forward in the design and fabrication of antennas for radio astronomy.

In the seventh chapter, authors Zhongmao Li, Mengjie Qin, Lanlan Liao, Xiaozhuang Hu, Pengzhan Liu, Xin Qiu, and Zhiqiang Li mention the latest developments in tunable microwave filters. The chapter covers various tunable bandpass filters based on different technologies such as on-chip, microstrip, and substrate-integrated waveguide. The chapter discusses the strengths and weaknesses of each technology by comparing the simulated and measured results on several engineering prototypes.

In the last chapter, author Veysel Demir presents a finite-difference time-domain (FDTD) code that was developed in the Julia programming language customized for efficient electromagnetic simulation of microstrip circuits. The code is tailored for faster simulation of microstrip circuits with lower memory requirements by employing an efficient form of FDTD updating equations, referred to as edge-length normalized formulation, as well as other simplifications that would make it sufficient to analyze single-layer microstrip circuits and produce scattering parameters. The chapter demonstrates how updating equations for fields, sources, and absorbing boundaries were modified to accommodate edge-length normalization, and how these equations were translated into a code in Julia. Finally, the chapter concludes with some examples of circuits simulated with the presented code.

The topic diversity, unique perspective, and rich content of each chapter make this book an enjoyable technology guide for academicians, graduate students, and engineers.

Hüseyin Şerif Savcı
Department of Electrical and Electronics Engineering,
İstanbul Medipol University,
İstanbul, Turkey

Chapter 1

Raw Kernels Packaged for Toasting: A Current Challenge for Microwave Technology

Nelly Lara

Abstract

The applications of microwaves for household use and processing at the industrial level have gained immense importance over traditional conduction or convection heating systems. Microwaves can heat and toast raw kernels such as peanuts, maize, hazelnuts, sunflower seeds, cocoa beans, coffee, and other food kernels with a notable decrease in energy consumption and heating time because this technology generates volumetric heating in each kernel with the absorption of the incident energy. However, for raw kernels packaged for toasting, the availability of a continuous or at least intermittent movement device is highly required in microwave ovens. This chapter will briefly discuss the basic principles available in microwave ovens based on the dielectric heating modes. Recent advances and challenges in uniform heating using movement or rotating devices. Also, the pros and cons of applying microwave energy in toasting processing operations will be discussed.

Keywords: microwave toasting, dry heating, microwave-packed kernels, household microwave ovens, first-generation snack

1. Introduction

Toasting is a unit operation that involves art and science, and this knowledge has grown from artisanal production to industrial processing of cereals and other dry kernels, with high levels of quality, efficiency, and hygiene. In toasting, sugars react with other carbohydrates and proteins to develop color and flavor compounds [1]. Thus, the toasting process of raw kernels is key to controlling the degree of modification of macro components and the development of aroma, texture, and color profiles of products, converted into edible forms. Specialty raw maize types subjected to traditional and microwave toasting are ready-to-eat kernels with soft and crunchy endosperm [2]. Raw almonds treated in a convective oven [3] and hot air-assisted radio frequency [4] present color, texture, and sensory attributes that meet consumer demands. Toasted peanuts, characterized by sensory analysis and aroma profiles, are critical data for the commercial peanut industry in selecting available cultivars to produce acceptable consumer whole-kernel snacks [5]. Cashew nut kernels toasted by hot air-assisted radio frequency enhance the nutritional quality and flavor [6]. In sunflower kernels, microwave toasting does not reduce the sensory quality of the

IntechOpen

product as it is expressed by overall consumer preference compared to convection toasting which increases kernels' hardness [7]. In chickpea snack production, toasting is a common process to improve their texture profile and palatability, so this topic has studies to compare the effect of different toasting methods [8]. On the other hand, toasting is a unit operation to achieve the characteristic flavor and texture in breakfast cereal and snack foods [9, 10]. In the processing of cocoa [11–13] and coffee [14–16], toasting develops the required profiles in color, aroma, and flavor sensations. In cereals and pseudocereals, toasting whole kernels could cause structural changes in starch and protein, breaking chains into partially soluble intermediate molecules with technological quality for the development of a variety of functional foods [17–19]. In legume flour, toasted chickpea kernels can be used in the development of various functional foods [20]. In addition, it has been found that toasting promotes the inhibition of anti-nutritional factors; e.g., phytic acid has decreased by 18% in quinoa toasted at around 180°C for 5 minutes [21]. Toasting is a traditional household process to convert raw seeds into edible kernels. The state of the art suggests that it is a current topic and has aroused great interest in researchers looking for innovation and new ingredients. Toasting is a unitary operation applied on a small, medium, and large scale through conventional and emerging heating systems based on conduction, convection, and radiation principles.

Referring to electromagnetic radiation, microwaves can toast dry raw materials under low temperatures (120–140°C) and short time compared with those of at least 150°C from an open flame, traditional toasting machines by long time and novel electric toasters [22]. Therefore, microwave-toasting household ovens and industrial systems can produce edible and healthy whole kernels in a short time. Hence, microwave ovens can be more efficient than conventional ovens and hot-air toasting technologies. However, most current applications are related to household microwaves with digital touchscreen control that only includes the heating program for popcorn, which does not require intensive movement of the packed kernel. In addition, the available studies aim at industrial toasting applications, where a microwave toasting machine has a feeding system that evenly distributes on the conveyor belt for raw materials like barley, oatmeal, whole kernels, seeds, pistachio, almonds, peanuts, and other nuts.

This chapter overviews recent research and innovation in microwave toasting for household and industrial applications. Also, the need for innovation in household microwave ovens to toast packaged kernels and obtain homemade snacks is addressed. Practical application approaches microwave toasting, and proposed packaging based on those available for microwave popcorn are also discussed. This chapter provides the first look at raw kernels packaged for toasting as a current challenge for microwave technology.

2. Overview on roasting and toasting foods

Dry heat is the heating medium for roasting and toasting methods in food production. Therefore, roasting and toasting raw food matrices do not use water or oil to convert them into ready-to-eat forms with special sensory attributes. Roasting is a culinary art applied to raw foods such as meat, fish, potatoes, plantains, apples, and other vegetables, that is, to raw food matrices with high moisture content (greater than 70%). Toasting, also of ancestral use, is applied to raw food matrices with low moisture content (less than 15%), such as cereals, other kernels, seeds, and dry nuts. Consequently, this chapter discusses the toasting of kernels in household microwave

DOI: http://dx.doi.org/10.5772/intechopen.1005169

ovens and the findings available to advance microwave applications in the food industry. **Figure 1a** and **b**, respectively, illustrate the increasing activity in toasting processes expressed as the total number of publications available worldwide and the

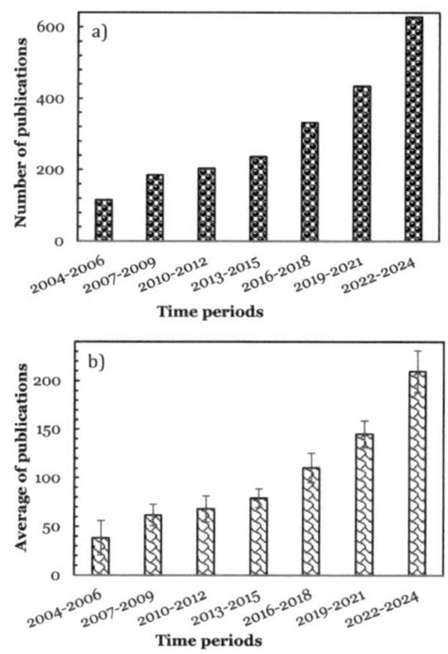

Figure 1.
Number and average of toasting research publications from 2004 to 2024: (a) the total number of publications available worldwide, and (b) an average of publications for three-year periods. Studies on popcorn and other food matrixes with intermediate and high moisture content are excluded.

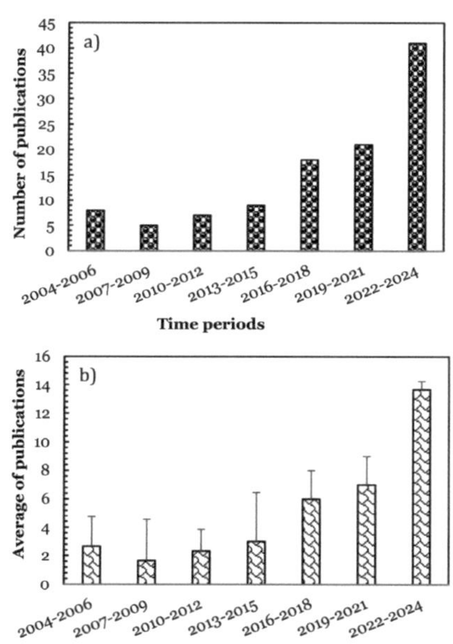

Figure 2.
Number and average of microwave toasting research publications from 2004 to 2024: (a) the total number of publications available worldwide, and (b) an average of publications for three-year periods. Studies on popcorn and other food matrixes with intermediate and high moisture content are excluded.

average number of publications for three-year periods collected from 2004 to 2024. Studies on popcorn or food materials with intermediate and high moisture content are excluded. Furthermore, interest in microwave toasting is also growing in the cycle evaluated, even though the number of publications and its increase is limited (**Figure 2a** and **b**). The general review of the results shows great progress in the last 6 years. It represents a challenge to the current popularity of microwave technology over conventional processes in other fields.

3. Raw kernels packaged for microwave toasting

3.1 Current research

Microwave toasting is a topic that has greatly interested researchers and toasting equipment developers worldwide. The knowledge generated shows the potential application of microwave toasting in the different fields of kernels, seed, and nut processing [22]. Recent studies reveal the feasibility of using a household microwave oven to toast maize kernels packed in kraft paper bags by 312–390 seconds. These results suggest that Andean maize types (*Zea mays amylacea* and *Zea mays saccharata*) with specialty toasting kernels can be packaged and subjected to dry heat in a microwave oven, alternating with rapid manual shaking during the toasting period [23, 24]. In these studies, the problem of non-uniform heating has been overcome by rapid manual shaking of the packages containing 150 g kernels. Microwave ovens widely used in homes and restaurants may be a good alternative to apply these findings in making ready-to-eat products by toasting other whole kernels, seeds, and nuts at home. However, no more information is available regarding raw kernels packaged for toasting in home microwave ovens, and the previous studies referred to are the first advances on this topic [2, 23–25]. The only cases are raw microwave maize types.

3.2 Raw microwave toasting maize types

Specialty maize types for toasting at home are floury maize (*Zea mays amylacea*) and sweet maize (*Zea mays saccharata*). However, maize kernel toasting using a pan is a slow process that can take 20 minutes to reach a uniform golden brown color with small kernel sizes [2]. Therefore, microwave toasting is an innovative application to prepare soft and crispy maize kernels in short heating-toasting times. Additionally, the packaged raw kernel for home microwave toasting aligns with the increasing interest in microwave toasting and the new trend toward healthy first-generation homemade snacks. **Figure 3** illustrates the ancestral form of toasting, the domestic practice of toasting at home, and the experimental trial of microwave-packed kernel proposed as an innovative alternative for processing floury and sweet maize types.

Floury maize used for toasting has kernels with completely soft endosperm, without vitreous fraction. The kernels are elongated in shape and large. The surface color varies from pale yellow to matte yellow, and the endosperm color is white. The toasted kernels are easy to chew, and the surface color changes to golden-brown with significant variations in physical, chemical, structural, and texture properties between raw and toasted kernels [2, 25]. **Figure 4a**, illustrates that elongated kernels, yellow pericarp, white, and entirely soft endosperm are the main characteristics of floury maize to make "Tostado." **Figure 4b**, shows the cross-section of maize kernel after toasting (ready-to-eat maize) with a soft appearance and easy to chew. The completely floury

DOI: http://dx.doi.org/10.5772/intechopen.1005169

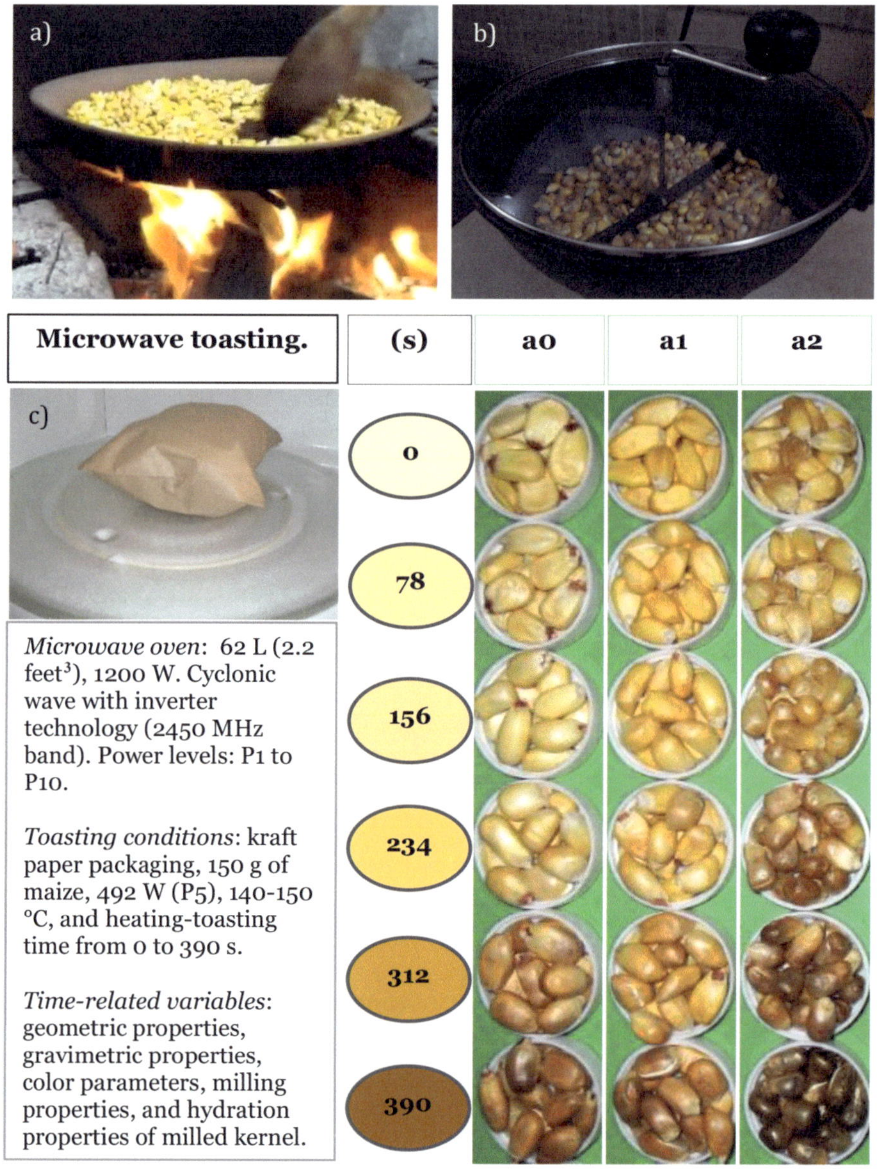

Figure 3.
Approach to maize kernel toasting from ancient to innovative: (a) ancestral toasting method (https://www.youtube.com/watch?v=IDJfAbAehKo), (b) domestic method, using an electric pan, and (c) experimental roasting test in a microwave oven.

endosperm appears as a simple starch-based material compared to starch-protein and lipid composites from other starchy sources. Therefore, **Figure 4**(**a** and **b**) suggests that these toasted whole kernels can be the first insight into future challenges to modify starch at the whole kernel endosperm level using microwave heating-toasting.

The mature native sweet maize kernels are shrunken and vitreous after drying. The vitreous fraction is translucent, and the color can vary between beige and yellow. The shrunken endosperm has a sugary-vitreous fraction more significant than the floury endosperm, with high total sugars and water-soluble index [24, 25]. The

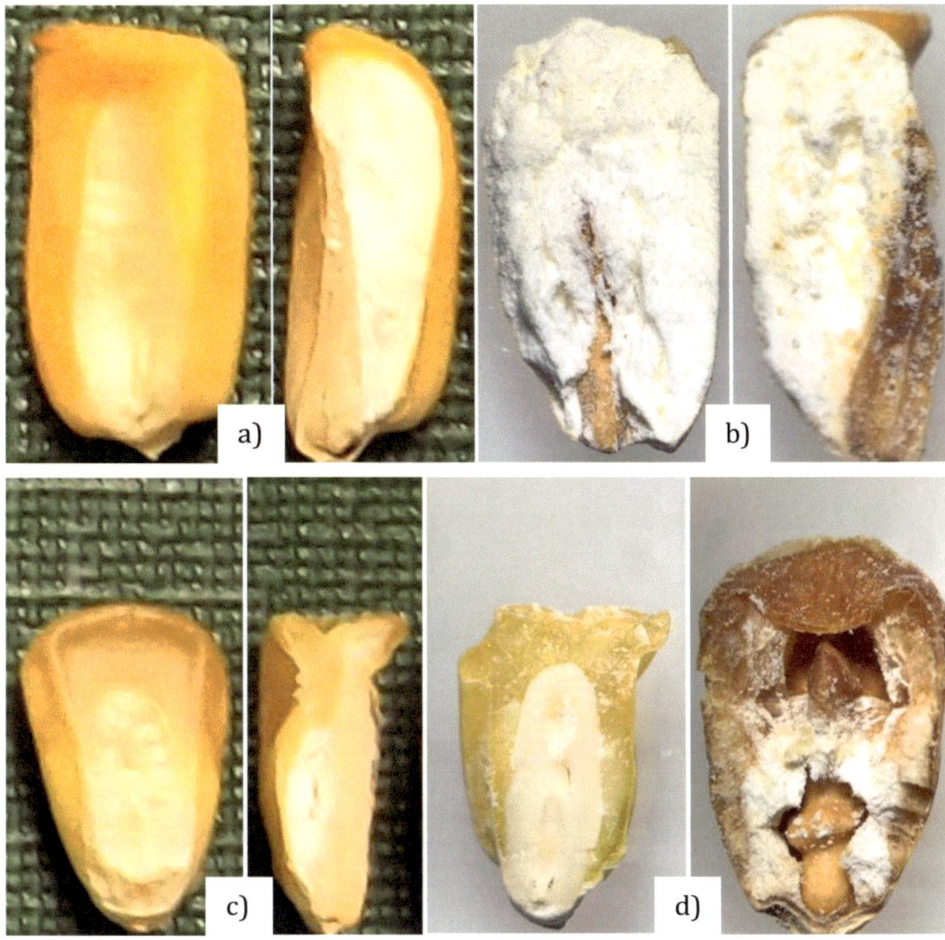

Figure 4.
Specialty maize kernels for toasting before and after being subjected to microwave heating at 492 W for 390 seconds, interspersing rapid manual movement of the packed kernel: (a) raw floury maize, (b) toasted floury maize, (c) raw native sweet maize, (d) toasted native sweet maize. Unpublished photographs.

high water-soluble solids non-converted in starch contribute to the developing crispy texture and more intense golden-brown color in sweet maize than floury maize during toasting [2]. **Figure 4c**, exhibits a shrunken yellow kernel and cross-section with a sugary-vitreous fraction greater than floury endosperm. **Figure 4d**, shows the cross-section of maize kernel after toasting (ready-to-eat maize) with a crispy texture to chew. There are significant structural changes between raw (**Figure 4c**) and toasted sweet kernels (**Figure 4d**), indicating that sweet maize toasted by microwave can be an interesting first-generation snack food.

According to the results of microwave heating-toasting at different times, most of the geometric, gravimetric, milling and color properties in maize kernels are time-related variables and correlate between them, reflecting the gradual effect of volumetric heating occurring in individual kernels, e.g., the decrease in density correlates with the increase in the volume of individual kernels due to water loss and kernel expansion with the formation of cavities in the endosperm [23]. **Tables 1** and **2** display the correlation coefficients established between the dependent variables used to characterize

DOI: http://dx.doi.org/10.5772/intechopen.105169

a1	Time	GMD	SA	IKV	BD	IP	APS	WSI	L*	ΔE*
Time		0.884	0.884	0.887	−0.909	0.873	−0.974	0.486	−0.933	0.877
GMD	0.884		0.999	0.953	−0.901	0.953	−0.918	0.539	−0.739	0.667
SA	0.884	1.000		0.999	−0.900	0.953	−0.917	0.541	−0.740	−0.740
KV	0.887	0.999	0.999		−0.894	0.946	−0.916	0.574	−0.748	0.671
BD	−0.909	−0.901	−0.900	−0.894		−0.905	0.931	−0.328	0.765	−0.718
IP	0.873	−0.901	0.953	0.946	−0.905		−0.917	0.366	−0.730	0.750
APS	−0.974	−0.918	−0.917	−0.916	0.931	−0.917		−0.427	0.881	−0.824
WSI	0.486	0.539	0.541	0.574	−0.328	0.366	−0.427		−0.532	0.382
L*	−0.933	−0.740	−0.748	−0.730	−0.739	−0.532	0.881	−0.532		0.765
ΔE*	0.877	0.668	0.671	0.750	0.667	0.382	−0.824	0.382	−0.938	

Table 1.
The correlation coefficients between some geometric, gravimetric, and color properties of floury maize a1 toasted by microwaves.

a2	Time	GMD	SA	IKV	BD	IP	APS	WSI	L*	ΔE*
Time		0.878	0.880	0.886	−0.903	0.863	−0.942	0.766	−0.997	0.990
GMD	0.878		1.000	0.999	−0.992	0.990	−0.933	0.830	−0.883	0.871
SA	0.880	1.000		1.000	−0.992	0.990	−0.934	0.831	−0.886	−0.886
KV	0.886	0.999	1.000		−0.992	0.991	−0.936	0.839	−0.892	0.880
BD	−0.903	−0.992	−0.992	−0.992		−0.977	0.946	−0.821	0.911	−0.903
IP	0.863	0.990	0.990	0.991	−0.977		−0.922	0.834	−0.863	0.846
APS	−0.942	−0.933	−0.934	−0.936	0.946	−0.922		−0.766	0.939	−0.932
WSI	0.766	0.830	0.831	0.839	−0.821	0.834	−0.766		−0.774	0.751
L*	−0.997	−0.883	−0.886	−0.892	0.911	−0.863	0.939	−0.774		−0.996
ΔE*	0.990	0.871	0.874	0.880	−0.903	0.846	−0.932	0.751	−0.996	

GMD: geometric mean diameter; SA: surface area; KV: kernel volume; BD: bulk density; IP: internal porosity; APS: average particle size; WSI: water-soluble index; L: lightness; ΔE*; overall color difference.*

Table 2.
The correlation coefficients between some geometric, gravimetric, and color properties of sweet maize a2 toasted by microwaves.

the microwave toasting of packaged maize kernels. Except for the water-soluble index (WSI) in floury maize, the remaining maize kernel properties are time-related variables. Their correlation coefficients indicate that the raw maize kernels packaged and subjected to microwave treatments mainly depend on geometrical properties because these dependent variables present a high correlation with gravimetric, milling, and color properties, for sweet maize with a smaller kernel size than that for floury maize. **Figure 5a** reveals that maize kernels' geometric mean diameter (GMD) is a time-related variable. Based on the characteristic size of each specialty maize for toasting, the GMD separates the microwave heating-toasting effect by type of maize. By plotting against the GMD, the bulk density (**Figure 5b**), the internal porosity of the whole kernels (**Figure 5c**), and the average particle size of milled kernels (**Figure 5d**) display the degree of existing correlation and reflect the influence on each material due to microwave treatment times. Regarding **Figure 6a**, the volume of maize

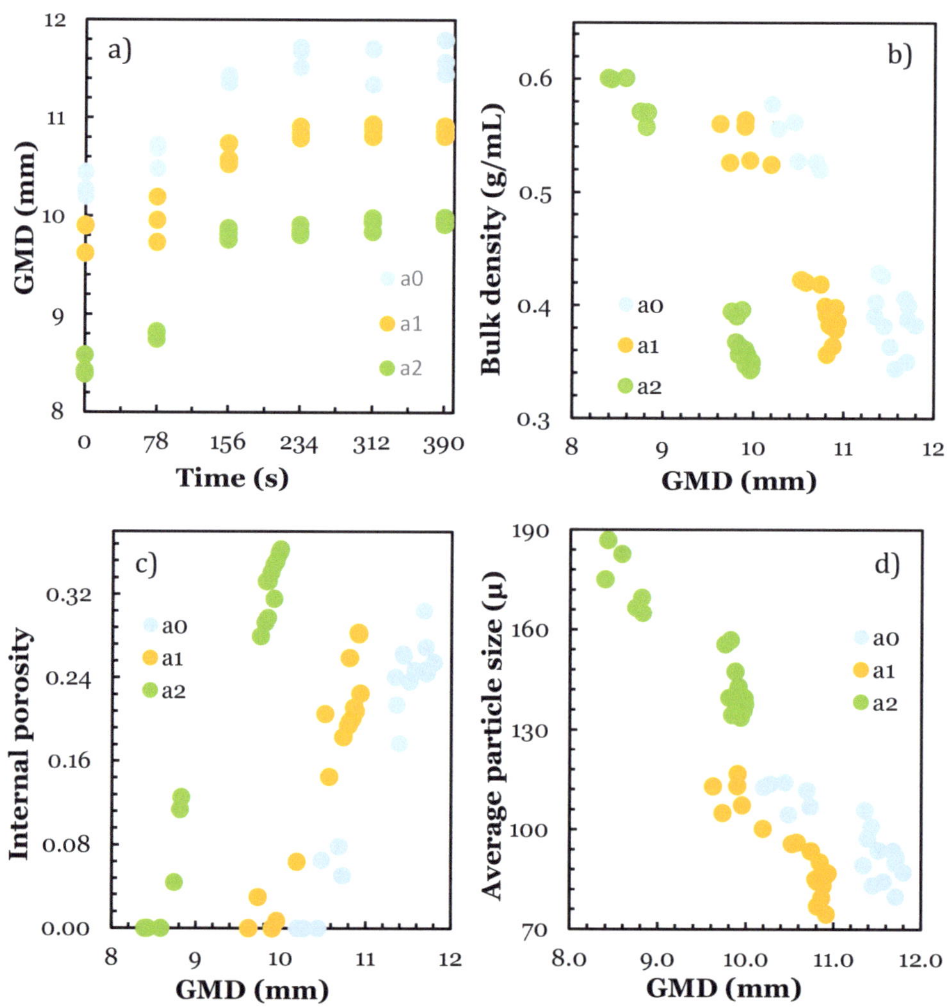

Figure 5.
Microwave heating-toasting effect on geometric mean diameter GMD and its correlation with other time-related variables determined in floury and sweet maize subjected to 492 W for 390 seconds: (a) GMD versus time, (b) bulk density versus GMD, (c) internal porosity versus GMD, and (d) average particle size versus GMD. Unpublished plots.

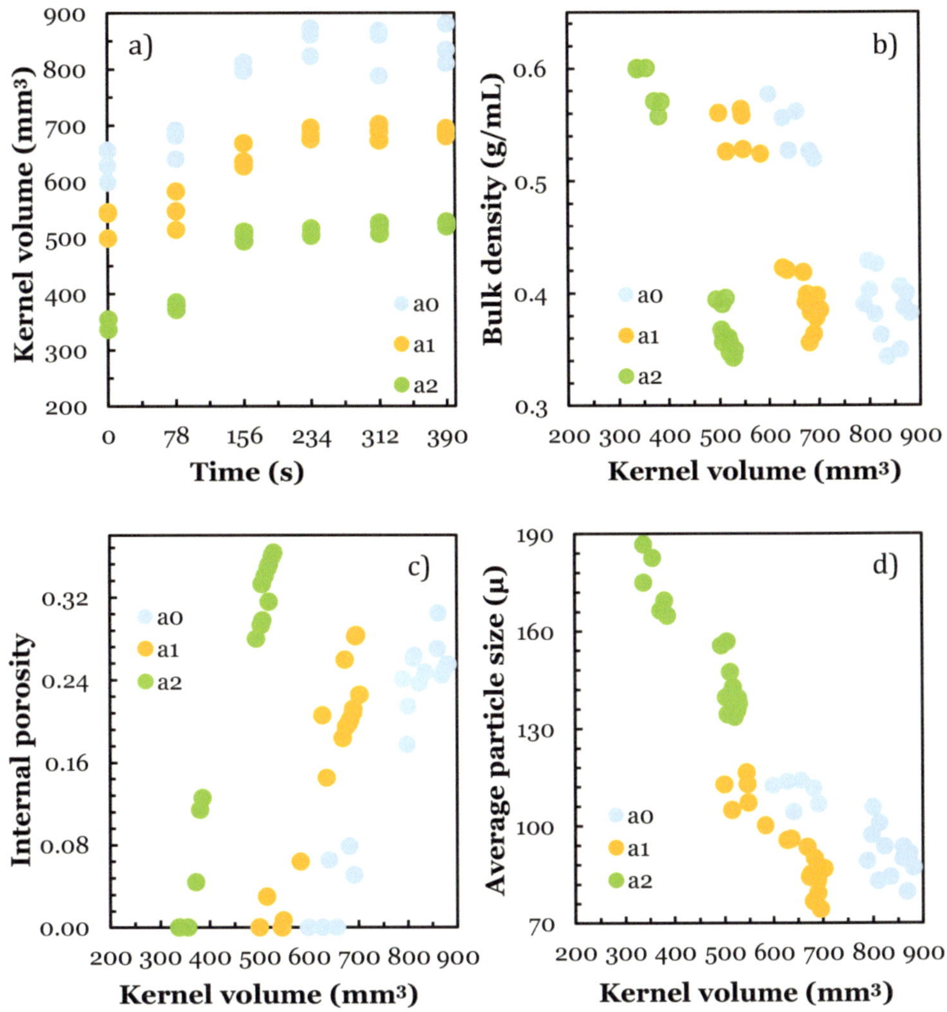

Figure 6.
Microwave heating-toasting effect on individual kernel volume and its correlation with other time-related variables determined in floury and sweet maize subjected to 492 W for 390 seconds: (a) kernel volume versus time, (b) bulk density versus kernel volume, (c) internal porosity versus kernel volume, and (d) average particle size versus kernel volume. Unpublished plots.

individual kernels varies by the effect of the microwave treatment time in agreement with the GMD. Thus, the kernel volume correlates with the bulk density, the internal porosity in whole kernels, and the average particle size of milled kernels, as shown in **Figure 6**(**b**, **c**, and **d**, respectively), demonstrating that maize geometry (size and shape) greatly influences the microwave heating of packaged kernels. **Figure 7a** shows that the bulk density is a variable related to the microwave treatment time. **Figure 7**(**b–d**) show the correlations with the variables associated with the structural properties of the endosperm in floury maize and sweet maize, expressed as internal porosity of whole kernels and average particle size in milled kernels. However, the variation does not contribute to differentiate between maize genotypes with floury endosperm, suggesting that the effect of microwave toasting may be similar for food matrices with the same structural pattern. **Figure 8a** exhibits the decreasing variation

DOI: http://dx.doi.org/10.5772/intechopen.1005169

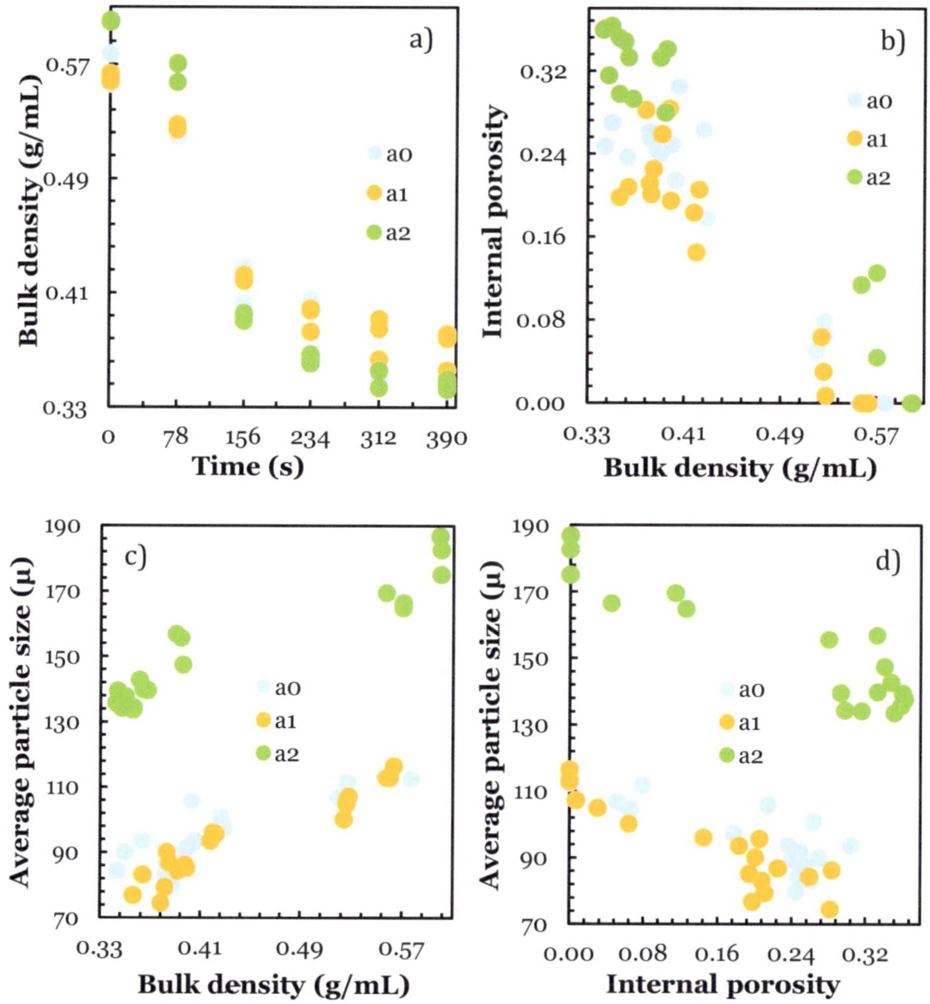

Figure 7.
Microwave heating-toasting effect on bulk density and its correlation with other direct and indirect time-related variables determined in floury and sweet maize subjected to 492 W for 390 seconds: (a) bulk density versus time, (b) internal porosity versus bulk density, (c) average particle size versus bulk density, and (d) average particle size versus internal porosity. Unpublished plots.

of the space color parameter L^* (lightness) by the effect of microwave heating time on the surface color of maize kernels. L^* parameter is a time-relate variable whose variation depends on the size of maize kernels. The correlation between the water-soluble index (WSI) and the L^* parameter shows the structural difference between sweet and floury maize (**Figure 8b**). However, the trends between the two types of floury maize are not well separated. A good correlation exists between the increasing overall difference in surface color (ΔE^*) and the decreasing L^* parameter (**Figure 8c**). This correlation suggests that rapid measurement of the L^* parameter can be considered for quality control of microwave toasting at the industrial level. The parameter L^*, relative to the surface ΔE^* due to the microwave treatment, correlates with the surface area of the maize kernel and exhibits a clear distinction in the trends of the three materials presented (**Figure 8d**). This is attributable to the differences in the size and shape of

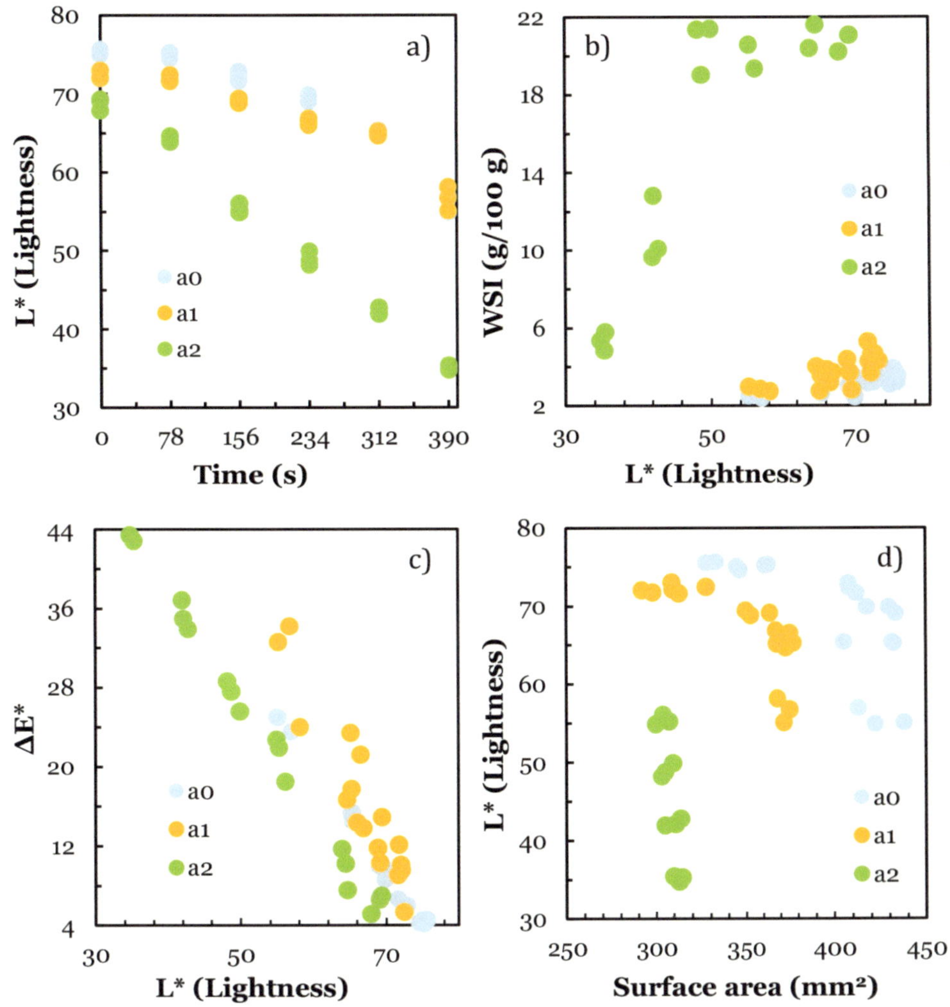

Figure 8.
Microwave heating-toasting effect on lightness (L) of the CIE color space and its correlations with other time-related variables determined in floury and sweet maize subjected to 492 W for 390 seconds: (a) L* versus time, (b) water-soluble index (WSI) versus L*, (c) color difference (ΔE*) versus L*, and (d) L* versus surface area. Unpublished plots.*

maize kernels, proving that the geometry of the product subjected to microwave treatment influences the color development of the maize surface area.

3.3 Practical application on raw microwave maize

All raw materials that, in the form of whole kernels, are transformed into first-generation snack foods by conventional toasting or frying methods can be packaged for toasting in a microwave oven, e.g., maize, chickpeas, beans, peanuts, and some nuts.

There are two practical applications to the types of maize. One is very recent with floury maize (*Zea mays amylacea*). Thus, microwave-packaged raw maize using floury maize is now available in the Ecuadorian consumer market. The product comes from a small-scale enterprise and is based on previous unpublished trials carried out in local research institutions. In this case, the major concern is the lack of a system for moving the packaging in the oven cavity that ensures continuous position changes of the kernels within the packaging during toasting. Raw popcorn (*Zea mays everta*) for household microwave ovens is another commercial product developed in the United States several decades ago. It is distinguishable by its endosperm matrix with natural popping capacity compared to the floury maize snack food (**Figure 9**). Popcorn is the primary first-generation homemade snack food compared to the same product obtained by conventional popping, reflecting the current popularity of microwave technology over traditional techniques.

3.4 Microwave packages for raw kernels

Many patents are related to microwave popcorn bags [26–28], so packaging refers to a bag with the basic structure for packaging raw popcorn kernels. A bag of this type is designed to unfold and increase its volume due to the steam generated during the toasting of the kernel. The free end vertices of the bag have a rounded configuration to prevent blocking with the oven cavity walls due to increased volume. The bag's volume depends

Figure 9.
Raw maize kernels packed for microwave oven: (a) microwaveable packaging with floury maize kernel, (b) microwaveable packaging with popcorn, and (c) first-generation snack foods made with floury maize and popcorn.

on the moisture content and the expansion capacity of the kernel. Therefore, the total width and the width of the folds are the parameters that can be varied in packaging for toasting other kernels with a reduced expansion volume, e.g., the reduced kernel expansion in floury maize compared with popping expansion in popcorn (**Figure 9**).

4. Innovation in microwave ovens for household and industrial use

Based on traditional toasting, the heat from the hot pan is transmitted evenly to the kernels through constant manual movement with a paddle. In this way, all kernels reach an approximately similar heating temperature. Consequently, continuous position changes of the kernels within the packaging are required during microwave toasting. Microwave technology in the toasting process aims to reduce operation time and improve final product quality. However, raw kernel packaged for microwave is challenging for household oven technology powered by the 2450 MHz frequency spectrum. The household microwave oven consists of a magnetron, a waveguide, and an aperture to a rectangular cavity. The microwave radiation, generated by magnetron produces reflection that leads to a nonuniform heating pattern even in standardized sample sizes [29]. According to the literature, the non-uniform distribution of the electromagnetic field due to the cavity geometry requires effective systems to achieve a uniform heating temperature [30], and attain adequate sample movement [31].

Likewise, a 915 MHz frequency is generated by a magnetron for industrial processes with a non-uniform electromagnetic field distribution. Therefore, the current studies aim to overcome this limitation. Solid-state generators have been recently introduced as capable of performing electronic control over the frequency and magnitude of the excited microwave field and heating processes for both continuous and pulse operations [29]. The narrow and stable bandwidth of the solid-state generator produces a single standing wave pattern, resulting in stable and predictable food heating patterns, inferring that the available information provides fundamental insights into solid-state microwave heating [32]. Considering the magnetron as a fundamental element in microwave technology, it is estimated that the increase in magnetrons in specific positions makes it possible to overcome the problem of non-uniform temperature distribution in the cavity [33]. Concerning the non-uniform electromagnetic field distribution caused by reflection at the cavity walls, its geometry seems to be decisive in how microwave heating is distributed, so sample rotational movement, cavity geometry, and frequency can affect temperature and electromagnetic field distribution at the industrial scale [30].

On the other hand, the advances in uniform microwave heating aim to improve the volumetric heating of solid samples with rotary motion devices adaptable to all kinds of microwave cavities irrespective of geometric shapes [34]. Therefore, microwave toasting suggests developing theoretical or empirical models related to this process. However, the assumptions required to apply Fick's second law of diffusion in drying processes do not satisfactorily describe the maize kernel toasting by microwave heating. In raw maize packaged for microwaves, there is influence from both the geometry of the packaging and the geometry of the individual kernels. In addition, microwave toasting with intermittent manual shaking causes an increase in maize kernel volume ascribed to internal porosity generated by water migration [2]. The toasting kinetics in raw microwave maize inside the packaging is a challenge for future understanding. The first insight is the wide range of variables related to microwave toasting time determined for microwave-packaged maize [23, 24]. The geometric shape of the product can also affect the reflection of microwave electromagnetic energy, power absorption efficiency, and penetration depth into the product from surface area [29]. The influence of sample size and geometry on the reflection of microwave electromagnetic radiation is still an unexplored field. The available information focuses on searching for optimal food geometries with high moisture content and specific sizes, such as potato tubers and portions of meat or fish. It is expected that the findings achieved will contribute to the design of packaged and unpackaged product geometries [35].

5. Microwave toasting for industrial application

Most current studies on kernel toasting apply the fundamental principles of microwave drying on a single layer of products spread on flat surfaces [36–39], so the drying irradiation fundament can be used in microwave industrial-scale toasting operations. The principle of microwave drying has generated a large amount of information worldwide. However, microwave toasting for industrial scaling requires further research, consistent with the growing trend of publications on microwave toasting observed in **Figure 2**, reported at the beginning. Consequently, this section highlights the following issues: (1) microwave toasting to make first-generation snack foods; (2) microwave expansion phase to obtain third-generation snack foods; (3) the toasting required in the cocoa and coffee industries; (4) the toasting of whole kernels, whether these are cereals, minor kernels or legumes, for the development of new ingredients rich in biocomponents or techno-functional properties; (5) inactivation of antinutritional compounds by toasting effect; and (6) the formation of acrylamide. **Table 3** summarizes the studies related to microwave technology with application perspectives in industrial kernel toasting.

Raw material	Higher goal	Microwave experiments	Relevant results	References
Sunflower seeds (*Helianthus annuus* L.)	To examine the effect of different toasting methods on sensory attributes, texture, and fat properties.	*Convection oven at 180 °C.* Time with four levels: 5, 10, 20, and 25 minutes. *Microwave oven, 2450 MHz* Power with three levels: 500, 600, 800 W. Time with three levels: 4, 6, and 8 minutes.	Microwave toasting did not diminish the sensory quality of the product.	[7]
Chickpeas	To determine the effect of different roasting methods on the compositions, powder properties, texture, and sensorial properties	*Conventional method.* Temperature at 250°C. *Microwave method 2450 MHz* Power with three levels: 100, 300, and 600 W. Time with three levels: 3, 5, and 7 min. *Microwave-assisted conventional method.* Microwave study treatment with additional toasting at 250°C.	The treatment to 300 W for 3 minutes and additional toasting at 250°C is reported for chickpea snack foods with the highest sensory scores.	[8]

Raw material	Higher goal	Microwave experiments	Relevant results	References
Peanuts, hazelnuts, walnuts	To improve the taste of the snack foods and results in energy with microwave toasting.	*Microwave oven 2450 MHz* Power in three levels: 500, 600, and 800 W. Time in three levels: 4, 6, and 8 min. *Convection oven at 180°C.* Time with four levels: 5, 10, 20, and 25 min.	Microwaving can ensure effective protection against hydrolytic rancidity and deterioration of primary oxidation compounds.	[7]
A base mixture of blue corn and powdered spinach reached with black bean	To evaluate the combined process of extrusion and microwave expansion on the third-generation snack food properties	*Extrusion.* Black beans in three levels: 11, 22, and 33%. *Microwave oven, 2450 MHz* Power of 1000 W for 23 s	It is possible to produce 3G snacks with bioactive properties and improved protein quality using a combination of extrusion cooking and microwave heating.	[40]
Corn flakes with 20%	To study the toasting of cornflakes by	*Microwave oven, 2450 MHz* Power in three levels: 504, 672, and 840 W.	The results suggest that microwave toasting could be a good alternative	[10]

Table 3.
Summary of the perspectives of microwave technology application in the kernel toasting industry.

5.1 Microwave toasting to make first-generation snack foods

Microwave toasting to obtain first-generation snack foods (1G) is a viable application in batch and continuous operations. The knowledge generated on the fundamentals of the food toasting process [22, 41] and the different experimental findings [7, 8, 42] have stimulated microwave oven developers to work on innovative designs with devices required in the experimentation laboratories and for the industrial scaling of microwave findings.

5.2 Microwave expansion phase to obtain third-generation snack foods

The microwave expansion phase to develop third-generation snack foods is a healthy way of improving nutrition value, texture properties, and sensory attributes [10, 43] . Thus, extruded, or pelleted products can be expanded indirectly by microwave heating-toasting phase in the snack food technology [43]. Third-generation snack foods are made with non-traditional ingredient formulations for typical extrusions. New or exotic ingredients can be incorporated to obtain products differentiated in nutritional value, texture properties, and sensory attributes [40]. However, the extrudates are intermediate products (pellets, flakes) that follow the next expansion phase by microwave heating-toasting, achieving fortification in nutritional value, new texture properties, and specific sensory attributes.

5.3 The toasting required in the cocoa and coffee industries

Toasting is an essential step in the cocoa industry that includes the combined use of time-temperature profiles to optimize the toasting degree. Toasting causes physicochemical and microstructural changes in cocoa beans that ensure their characteristic aromas and flavors. A comprehensive review of the state of the art of cocoa toasting provides important insights into the use of microwaves in cocoa toasting [11], and allows consider the microwave technology to toast cocoa beans for future studies. The impact of microwave-toasting cocoa beans on the aroma and phytochemical profile of dark chocolates (70% cocoa) is reported compared to chocolates produced from cocoa beans toasted by convection at 130°C for 30 minutes. Indeed, the results validate the use of microwave toasting for cocoa beans and project as a promising technological step that can be scaled with further research to improve the specialty sensory characteristics in chocolate production [11, 44].

Coffee is a hot drink consumed worldwide. It is made from toasted coffee beans, and its characteristic aroma and flavor depend on specialty compounds developed during the toasting step [45]. Thus, toasting is one of the critical steps for the processing of coffee beans to be used in beverages. There are limited studies on coffee bean toasting by microwaves. However, the comparative results between conventional and microwave toasting reveal a similar effect on volatile components and less caffeine and chlorogenic acid reduction with microwave toasting [46], so microwave radiation can be used as an innovative toasting technology for coffee beans.

5.4 The toasting of whole kernels for the development of new ingredients

The heating and toasting of whole kernels are topics of current research that seek to establish specific conditions to modulate changes in the multi-component matrix of each type of kernel [17]. Cereals [36], minor kernels [19], legumes [8, 47], oilseeds [48, 49], and other seeds can be potential sources of new food ingredients through microwave heating and toasting. The studies carried out in recent years show the effect of microwave technology on the microstructure, composition, bioactive components, and physicochemical properties of different kernels [47–50].

5.5 Inactivation of antinutritional compounds by toasting effect

Some starchy kernels modified (treated) with dry heat can exhibit specific physiological functions such as antidiabetic and anti-inflammatory activity from residues encapsulated in starch-protein structures [17]. For some legumes (pulses), the thermal treatment contributes to deactivating antinutritional factors such as tannin, protease inhibitors (trypsin, chymotrypsin), and phytic acid, while preserving the functional properties of legumes. In summary, these works provide data support to investigate microwave use to deactivate antinutritional factors in raw materials [21, 51].

5.6 Acrylamide formation

Acrylamide is classified as a potential carcinogen (Group 2A) in humans. However, little is still known about the levels of acrylamide in different foods and their potential risk [52, 53]. The major pathway of acrylamide formation seems to be the Maillard reaction, in particular, in the presence of the amino acid asparagine [54]. Therefore, the acrylamide content in cereals depends on several factors, such as the level of free asparagine, the composition of the food matrix, and the time-temperature profiles of the thermal treatments [52]. Toasting is important in the nut industry to develop flavor and texture. However, the Maillard reaction may result in acrylamide formation. For example, raw and salted pistachios treated under hot air, infrared, and microwave toasting present the highest amounts of acrylamide in raw pistachios toasted by infrared. In contrast, a lower amount is observed for the microwave irradiation [55]. Consequently, research on opportunities for the application of microwave toasting technology should have acrylamide content as another measurement variable.

6. Conclusion and future prospective

This chapter shows that toasting kernels, seeds, and nuts is a current topic and has sparked interest worldwide, focusing on snack foods and developing ingredients with new properties. Except for popcorn and its natural popping capacity, microwave toasting of packaged whole kernels is an unexplored field. Its importance is based on the opportunity to prepare different types of first-generation snack foods at home. Based on the popularity achieved by the microwave oven for home use and the acceptance of first-generation snack food among consumers, innovation in the microwave oven for home use is a challenge for microwave technology. The lack of uniformity in volumetric microwave heating due to the reflection of electromagnetic radiation is recognized in domestic ovens and conveyor belts for industrial applications. Uniform temperature distribution in packaged and unpacked food and the heating space is the highest goal of the advances. The experiments could be designed with independent and dependent variables based on the source of microwave generation, the microwave heating space specifications, and the food's properties, including type, shape, and size.

Acknowledgements

My gratitude goes to the Vice-Rectorate of Research, Doctorates, and Innovation of the Central University of Ecuador.

Conflict of interest

The authors declare no conflict of interest.

Author details

Nelly Lara
Central University of Ecuador, Quito, Ecuador

*Address all correspondence to: nvlara@uce.edu.ec

References

[1] Breslin JC, Knott K. 15 - toasting. In: Perdon AA, Schonauer SL, Poutanen KS, editors. Breakfast Cereals and How they Are Made. 3rd ed. Oxford: AACC International Press; 2020. pp. 299-321

[2] Lara N, Ruales J. Physical and hydration properties of specialty floury and sweet maize kernels subjected to pan and microwave toasting. Journal of Cereal Science. 2021;**101**:103298

[3] Lipan L, Cano-Lamadrid M, Vázquez-Araújo L, Łyczko J, Moriana A, Hernández F, et al. Optimization of roasting conditions in hydroSOStainable almonds using volatile and descriptive sensory profiles and consumer acceptance. Journal of Food Science. 2020;**85**:3969-3980

[4] Xu Y, Liao M, Wang D, Jiao S. Physicochemical quality and volatile flavor compounds of hot air-assisted radio frequency roasted almonds. Journal of Food Processing and Preservation. 2020;**44**:e14376

[5] Dean LL, Campbell RA, Stoner-Harris T, Hung Y-C, Hendrix KW, Adhikari K. Profiling seventeen cultivars of roasted peanuts by descriptive sensory and flavor volatile analyses. Measurement: Food. 2023;**11**:100105

[6] Liao M, Zhao YH, Xu Y, Gong C, Jiao S. Effects of hot air-assisted radio frequency roasting on nutritional quality and aroma composition of cashew nut kernels. LWT-Food Science and Technology. 2019;**116**:108551

[7] Goszkiewicz A, Kołodziejczyk E, Ratajczyk F. Comparison of microwave and convection method of roasting sunflower seeds and its effect on sensory quality, texture and physicochemical characteristics. Food Structure. 2020;**25**:100144

[8] Tekgül Barut Y, Çalışkan Koç G, Rayman Ergün A, Bozkır H, Pandiselvam R. Effect of different roasting methods on the proximate by composition, flow properties, amino acid compositions, colour, texture, and sensory profile of the chickpeas. International Journal of Food Science & Technology. 2023;**58**:482-492

[9] Bhattacharya S. Chapter 12 - processes and machinery. In: Bhattacharya S, editor. Snack Foods. London: Academic Press; 2023. pp. 321-382

[10] González LC, Loubes MA, Tolaba MP. Evaluation of microwave toasting of corn flakes. Journal of Food Processing and Preservation. 2018;**42**:e13671

[11] Peña-Correa RF, Ataç Mogol B, Fogliano V. The impact of roasting on cocoa quality parameters. Critical Reviews in Food Science and Nutrition. 2022:1-14. Published online 16 November 2022 [In press]

[12] Castro-Alayo EM, Idrogo-Vasquez G, Siche R, Cardenas-Toro FP. Formation of aromatic compounds precursors during fermentation of criollo and Forastero cocoa. Heliyon. 2019;**5**:e01157

[13] Velasquez-Reyes D, Rodriguez-Campos J, Avendano-Arrazate C, Gschaedler A, Alcazar-Valle M, Lugo-Cervantes E. Forastero and criollo cocoa beans, differences on the profile of volatile and non-volatile compounds in the process from fermentation to liquor. Heliyon. 2023;**9**:e15129

[14] Flores-Chávez B, Garza-López PM, Angel-Cuapio A, Hernández-León S,

Espinosa-Zaragoza S, Villafaña-Rivera FJ, et al. Analysis verbal sensory methods in the evaluation of organic coffee with different roasting intensities. International Journal of Food Properties. 2022;**25**:708-716

[15] Palma FD, Iacono F, Toffanin C, Ziccardi A, Magni L. Scalable model for industrial coffee roasting chamber. Procedia Computer Science. 2021;**180**:122-131

[16] Liao Y-C, Kim T, Silva JL, Hu W-Y, Chen B-Y. Effects of roasting degrees on phenolic compounds and antioxidant activity in coffee beans from different geographic origins. LWT-Food Science and Technology. 2022;**168**:113965

[17] Li L, Wang Q, Liu C, Hong J, Zheng X. Effect of oven roasting on major chemical components in cereals and its modulation on flour-based products quality. Journal Food Science. 2023;**88**:2740-2757

[18] Fukui M, Islam MZ, Lai H-M, Kitamura Y, Kokawa M. Effects of roasting on storage degradability and processing suitability of brown rice powder. LWT-Food Science and Technology. 2022;**161**:113277

[19] Kheto A, Joseph D, Islam MZ, Dhua S, Das R, Kumar Y, et al. Microwave roasting induced structural, morphological, antioxidant, and functional attributes of quinoa (*Chenopodium quinoa* Willd). Journal of Food Processing and Preservation. 2022;**46**:e16595

[20] Kaur R, Prasad K. Process optimization for the development of traditionally roasted chickpea flour for meal replacement beverages. Food Chemistry Advances. 2023;**3**:100452

[21] Sharma S, Kataria A, Singh B. Effect of thermal processing on the bioactive compounds, antioxidative, antinutritional and functional characteristics of quinoa (*Chenopodium quinoa*). LWT-Food Science and Technology. 2022;**160**:113256

[22] Das PP, Duarah P, Purkait MK. 5 - fundamentals of food roasting process. In: Jafari SM, editor. High-Temperature Processing of Food Products. Cambridge: Woodhead Publishing; 2023. pp. 103-130

[23] Lara N, Osorio F, Ruales J. Variables related to microwave heating-toasting time and water migration assessment with kernel size approaches of specialty maize types. Journal of the Science of Food and Agriculture. 2022;**102**:6088-6099

[24] Lara N, Portilla A, Osorio F, Ruales J. Modeling of the microwave heating-toasting time-related variables and characterization of non-isothermal rheological properties of floury and sweet specialty maize kernels. Applied Food Research. 2022;**2**:100119

[25] Lara N, Vizuete K, Debut A, Chango I, Campaña O, Villacrés E, et al. Underutilized maize kernels (*Zea mays* L. var. *Amylacea* and var. *saccharata*) subjected to pan and microwave toasting: A comparative structure study in the whole kernel. Journal of Cereal Science. 2021;**100**:103249

[26] Almog Y. Method and Package for Evenly Roasting Seeds in a Microwave, France. European Patent No. EP 3 861 832 B1. Paris, France: Europen Patent Office; 2021

[27] France DW. Microwave Popcorn Bag, Canada. Google Pantents No.CA3091430A1. 2019

[28] Galceran Martorell C. Bag for Cooking Corn Grains in a Microwave, United States. U.S. Patent No. US20190009971A1. Washington, DC: U.S. Patent and Trademark Office; 2019

[29] Koutchma T. Chapter 2 - heating characteristics of microwave systems and dielectric properties of foods. In: Koutchma T, editor. Microwave and Radio Frequency Heating in Food and Beverages. London: Academic Press; 2023. pp. 29-53

[30] Altin O, Skipnes D, Skåra T, Erdogdu F. A computational study for the effects of sample movement and cavity geometry in industrial scale continuous microwave systems during heating and thawing processes. Innovative Food Science & Emerging Technologies. 2022;77:102953

[31] Ye J, Lan J, Xia Y, Yang Y, Zhu H, Huang K. An approach for simulating the microwave heating process with a slow- rotating sample and a fast-rotating mode stirrer. International Journal of Heat and Mass Transfer. 2019;**140**:440-452

[32] Zhou X, Tang Z, Pedrow PD, Sablani SS, Tang J. Microwave heating based on solid-state generators: New insights into heating pattern, uniformity, and energy absorption in foods. Journal of Food Engineering. 2023;**357**:111650

[33] Salvador AA, Teleken JT, Travassos XL, Avila SL, Carciofi BAM. Multiphysics modeling to assist microwave cavity design for food processing. Journal of Electrical and Computer Engineering. 2022;**2**:1-10

[34] Ye J, Xia Y, Yi Q, Zhu H, Yang Y, Huang K, et al. Multiphysics modeling of microwave heating of solid samples in rotary lifting motion in a rectangular multi-mode cavity. Innovative Food Science & Emerging Technologies. 2021;**73**:102767

[35] Deng A, Li X, Qiu W, Li L, Tao N, Jin Y. Geometry optimization of microwavable food to improve heating uniformity. Journal of Food Engineering. 2024;**369**:111945

[36] An N-N, Li D, Wang L-J, Wang Y. Microwave irradiation of corn kernels: Effects on structural, thermal, functional and rheological properties of corn flour. Food Hydrocolloids. 2023;**143**:108939

[37] Arballo JR, Campañone LA. 7 - application of microwave processing in the food industry. In: Jafari SM, editor. Emerging Thermal Processes in the Food Industry. Cambridge: Woodhead Publishing; 2023. pp. 165-226

[38] Gaikwad PS, Sunil CK, Negi A, Pare A. Effect of microwave assisted hot-air drying temperatures on drying kinetics of dried black gram papad (Indian snack food). Applied Food Research. 2022;**2**:100144

[39] Shao Z, Song Y, Hong Y, Tao S, Sun J, Liu C, et al. The extension of vacuum microwave drying time improved the physicochemical properties, in vitro digestibility and antioxidant activity of brown rice flour. LWT-Food Science and Technology. 2023;**184**:115023

[40] Neder-Suárez D, Vázquez-Rodríguez JA, González-Martínez BE, Meléndez-PizarroCO,Hernández-OchoaLR, Murowaniecki-Otero D, et al. Effect of using alternative flours on the development and characteristics of a third-generation snacks. Food Chemistry Advances. 2024;**4**:100571

[41] Kutlu N, Pandiselvam R, Saka I, Kamiloglu A, Sahni P, Kothakota A. Impact of different microwave treatments on food texture. Journal of Texture Studies. 2022;**53**:709-736

[42] Goszkiewicz A, Kochanska E, Korczak K, Potapov VO, Prasol S. Influence of microwave treatment on

quality parameters of snacks food. Impact issuees. Acta Innovations. 2020;**36**:64-80

[43] Panak Balentic J, Babic J, Jozinovic A, Ackar D, Milicevic B, Muhamedbegovic B, et al. Production of third-generation snacks. Croatian Journal of Food Science and Technology. 2018;**10**:98-105

[44] Lemarcq V, Monterde V, Tuenter E, Walle DVD, Pieters L, Sioriki E, et al. Flavor diversification of dark chocolate produced through microwave roasting of cocoa beans. LWT-Food Science and Technology. 2022;**159**:113198

[45] Ganju E, Chawla K, Yang S, Chawla N. Time-resolved roasting-induced microstructural evolution of arabica coffee beans from Brazil, Colombia, and Ethiopia captured using x-ray computed micro-tomography. Journal of Food Engineering. 2024;**361**:111733

[46] Yoon S, Jeong H, Jo SM, Hong SJ, Kim YJ, Kim JK, et al. Chemosensoric approach for microwave- or oven-roasted *Coffea arabica* L. (cv. *Yellow Bourbon*) using electronic sensors. LWT-Food Science and Technology. 2022;**167**:113844

[47] Najib T, Heydari MM, Tu K, Meda V. Modification in starch structure of soaked and germinated lentil seeds under various thermal processing methods, including conventional, microwave, and microwave-assisted techniques. Food Chemistry Advances. 2023;**2**:100267

[48] Yin WT, Yang CJ, He XY, Zhao YH, Liu HM, Zhai ZQ, et al. Comparison of microwave and hot-air roasting on microstructure of sesame seed, aroma-active, hazardous components, and sensory perception of sesame oil. Food Chemistry X. 2023;**20**:101045

[49] Yin W-T, Shi R, Li K, Wang X-D, Wang A-N, Zhao Y-H, et al. Effect of microwave pretreatment of sunflower kernels on the aroma-active composition, sensory quality, lipid oxidation, tocopherols, heterocyclic amines and polycyclic aromatic hydrocarbons of sunflower oil. LWT-Food Science and Technology. 2022;**170**:114077

[50] Özcan MM, Uslu N. The effect of oven and microwave roasting on bioactive properties, phenolic components and fatty acid compositions of soybean and peanut seeds. Food and Humanity. 2024;**2**:100250

[51] Manikpuri S, Kheto A, Sehrawat R, Gul K, Routray W, Kumar L. Microwave irradiation of guar seed flour: Effect on anti-nutritional factors, phytochemicals, in vitro protein digestibility, thermo-pasting, structural, and functional attributes. Journal of Food Science. 2024;**n/a**

[52] Mesias M, Delgado-Andrade C, Morales FJ. An updated view of acrylamide in cereal products. Current Opinion in Food Science. 2022;**46**:100847

[53] Žilić S. Chapter 10 - acrylamide in soybean products, roasted nuts, and dried fruits. In: Gökmen V, Mogol BA, editors. Acrylamide in Food. 2nd ed. London: Academic Press; 2024. pp. 201-222

[54] Koutchma T. Chapter 4 - Microwave heating and quality of food. In: Koutchma T, editor. Microwave and Radio Frequency Heating in Food and Beverages. London: Academic Press; 2023. pp. 81-111

[55] Asadi S, Aalami M, Shoeibi S, Kashaninejad M, Ghorbani M, Delavar M. Effects of different roasting methods on formation of acrylamide in pistachio. Food Science & Nutrition. 2020;**8**:2875-2881

Chapter 2

Microwave Rewarming of Limb Frostbite Injuries (Simulation of Modes)

Grigory Dunaevskiy, Alexander Nechaev, Polina Smygalina and Igor Dorofeev

Abstract

Cold limb injury remains a serious and widespread condition both in countries with a cold climate and in regions located close to the equator, but having high-altitude territories. There is no medical equipment for the treatment of this condition. One of the ways to solve this problem is the use of microwave radiation at an early stage, which penetrates deep into the cooled volume and, accordingly, can accelerate the activation of internal vessels. The technical implementation of this approach involves the selection of the appropriate frequency and power of radiation, as well as the creation of a microwave chamber in which it is possible to ensure a sufficiently uniform heating of the entire volume of the frostbitten limb. The results of modeling the distributions of electromagnetic and thermal fields in the volume of the heated limb at microwave frequencies allowed for medical applications are presented.

Keywords: limb frostbite, microwave heating, field modeling, temperature distribution, amputation prevention

1. Introduction

A large part of the world population lives in areas with subzero winter temperatures, where frostbites of extremities are quite common. Such injuries occur in warmer climates as well—in high-mountain regions, high-altitude stations and ski resorts. Frostbite of hands or feet develops unnoticeably and without pain, but the complications can be extremely severe and often result in amputations or disability [1–4].

Surprisingly, to date, no equipment exists to treat this condition.

The aforesaid negative outcomes tend to develop with deep, volume frostbites when due to low temperatures, blood flow and lymph circulation are impaired or entirely stop in the limb volume. In these cases, any attempts to rewarm outer areas (the most common scenario in frostbites) result in a rapid dilatation of topmost vessels, which try to "push" their contents. At the same time, main deep-seated vessels remain ischemic due to the cold and prevent the blood/lymph flow, which invariably causes thrombosis, ruptures, and necrosis. This explains why fingers and toes are particularly affected. They rewarm quickly but amputations are practically inevitable if other vessels of the extremities (palms, forearms, feet, lower legs) are not activated.

One of the possible approaches to this problem is the use of deeply penetrating, microwave radiation, which helps rewarm the inner volume of an extremity and activate innermost blood vessels more quickly.

In medicine, electromagnetic fields (EMFs) have wide applications, including deep rewarming. One of the uses is physical therapy [5–8]. Physical therapy machines are comprised of a microwave generator and a set of antennas, which are applicators that focus microwave radiation on the specific area of the body. The maximum generator power is 200 W [9] up to 400 W [10] and both permanent and pulsed modes are possible. During the procedure, the patient wears safety goggles and the technician must leave the room. Such machines are ineffective for rewarming of frostbitten limbs because they do not rewarm the entire volume of the limb simultaneously and uniformly.

Another EMF application in medicine is induction of hyperthermia in malignant tumors [11–13]. For this purpose, more powerful electromagnetic machines, also with narrowly targeted antennas are used. Thus, this equipment is not applicable for the condition in question either.

In this paper, we describe the method of limb rewarming in a closed metal chamber, which provides thawing of the affected limb simultaneously from all sides [14]. The patient and the technician can be completely protected from the negative exposure. Moreover, to achieve the desirable effect, considerably less powerful microwave radiation is sufficient.

I. Gorelik demonstrated the efficacy of this technique in rabbits [15].

In **Figure 1**, scintigrams of rabbit limbs following frostbite and a recovery procedure with microwaves are shown.

Clearly, more research is in order to apply this technique to treatment in humans.

We will not explore here the medical aspects of this technique, assuming that all positive and negative sides of microwave rewarming were investigated at the time of designing of these physical therapy and hyperthermia machines. However, it is important to note that to ensure the activity of blood and lymphatic vessels, the entire

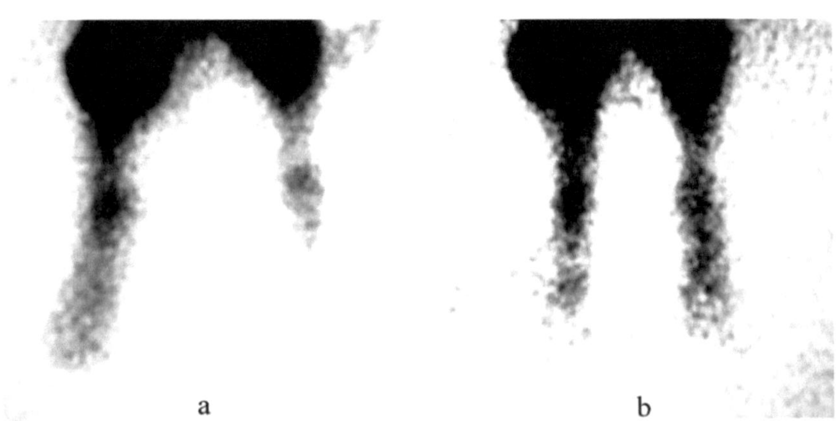

Figure 1.
Scintigrams of rabbit limb vessels following frostbite and microwave rewarming: 4 hours after frostbite (a), 1 hour after microwave exposure (b).

DOI: http://dx.doi.org/10.5772/intechopen.1002216

volume of a frostbitten limb must be rewarmed to over 15°C [4] as this is the absolute threshold when the vessels do not simply "open" but create the necessary flow of fluids.

For an effective and safe application of the proposed technique in humans, it is important to examine it as carefully as possible, using mathematical and numerical modeling as well as physical modeling on limb-equivalent dielectric phantoms. The modeling made it possible to better describe field distribution within the volume of an affected limb and experiments in the phantoms resulted in recommended values of the microwave power fed to the chamber as well as the duration of rewarming procedures.

Clearly, the reader is interested in practical applications of the technique. Normally, rigorous healthcare regulations require meticulous preclinical and clinical research of new treatment methods. Because for the actual application of the technique, a certified SMVI-200 physiotherapeutic device was used and the microwave radiation was far less powerful than the maximum allowed for this device, the TSU Committee of Bioethics allowed the team of developers to enroll informed volunteers in the studies. The final part of the chapter gives an overview of the findings.

2. Microwave rewarming in the chamber. Principle, features, physical problems

A multi-wave metal chamber was proposed as the basis of the recovery technique when the electromagnetic microwave field is fed to an affected limb from all sides simultaneously. A schematic diagram of this device is shown in **Figure 2**.

The limb is placed inside a chamber (1) through the opening in one of the chamber walls. The opening is equipped with a radioprotective sleeve (2). The excitation is performed by a generator (3) from the outside of the chamber via the built-in antenna (4).

Figure 3 shows one of the possible designs of this device, where a generating unit of a SMVI-200 medical physiotherapeutic device [16] is used as a microwave generator. This generating unit produces microwave frequency of 2.45 GHz. Notably, even though the highest power of this device is 200 W, to achieve the desired effect, 30–40 W of power output is sufficient.

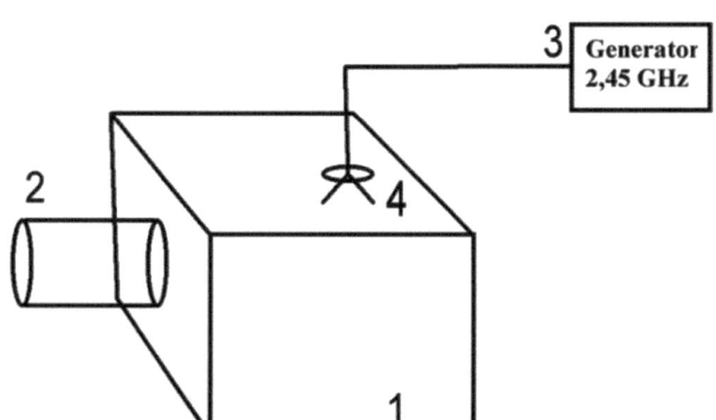

Figure 2.
Schematic diagram of the microwave rewarming device: 1 – Microwave chamber, 2 – Safety sleeve, 3 – Microwave generator, 4 – Source antenna [14].

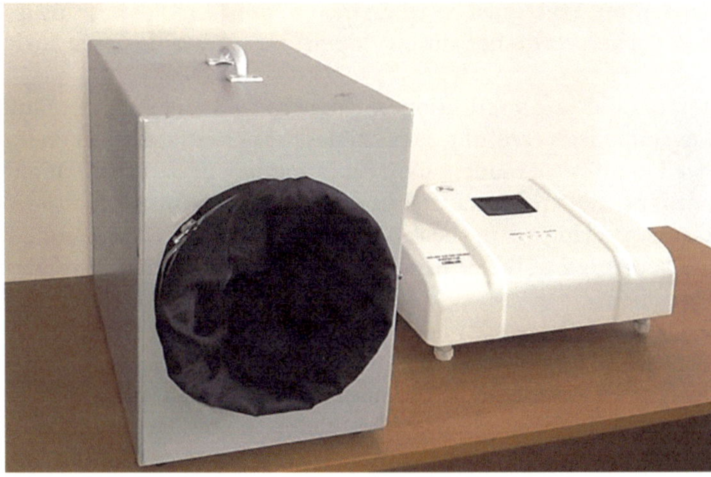

Figure 3.
A design of the rewarming limb device with a microwave chamber and a generating unit of SMVI-200 medical physiotherapeutic device [16].

If we assume that microwave rewarming is sufficiently explored from the medical viewpoint, the application of this design to treat humans, though seemingly simple, should be justified clearly in terms of technology.

The key task is to determine the distribution of electromagnetic and thermal fields in the chamber and in the volume of the affected limb and estimate the extent of nonuniformity of the fields and their behavior in fingers and toes. Palms, fingers, and toes require special attention because they can be considered critical zones because on the one hand, they are most susceptible to frostbite, and on the other, they can be rewarmed with microwave radiation far faster than the remaining limb volume, which may negatively affect treatment outcomes.

The choice of the microwave frequency is highly important. Biological tissues that form limbs (skin and fat, muscles, bones) are considerably different in terms of dielectric permittivity and electrical conductivity, which vary at different frequencies. Thus, the rewarming depth and rapidness depend on the frequency.

It is also important to study the behavior of the rate of reflection of microwaves off the chamber and related with it voltage standing wave ratio (VSWR), which determines what amount of the power fed by the generator reaches the chamber and gets absorbed inside the limb and what amount is reflected back and not used for rewarming.

3. Field modeling in a dielectric phantom

A human limb (an arm or a leg) is a multi-component object of a complex and continuously varying shape, which is quite difficult to describe with a mathematical model. Even a greater difficulty is to create a physical phantom that accurately imitates a limb with all its variety of tissues. For this reason, both mathematical models and physical phantoms are inevitably simpler yet still suitable for predicting rewarming dynamics of a real object.

DOI: http://dx.doi.org/10.5772/intechopen.1002216

The mathematical problem of describing electromagnetic and thermal fields in the limb placed inside a microwave chamber can be presented as a simultaneous solution of Maxwell equations describing the electromagnetic fields,

$$\begin{cases} rot\vec{H} = -i\omega\varepsilon\varepsilon_0\vec{E} + \sigma\vec{E} \\ rot\vec{E} = i\omega\mu\mu_0\vec{H} \end{cases}, \tag{1}$$

and a Pennes bio-heat equation describing the process of heating a limb (or an imitating phantom) with the electromagnetic field [17]:

$$\rho C\frac{dT}{dt} = k\nabla T\left(\vec{r},t\right) - \omega_B C_B \rho_B (T - T_B) + q_M\left(\vec{r},t\right) + q_{MW}\left(\vec{r},t\right). \tag{2}$$

This equation consists of the following terms: $-\omega_B C_B \rho_B (T - T_B)$ is a summand arising from the heat transfer by the blood flow, $q_M\left(\vec{r},t\right)$ is energy produced in the tissue as a result of metabolism [W/m^3], $q_{MW}\left(\vec{r},t\right)$ is energy absorbed in the tissue due to microwave electromagnetic fields [W/m^3].

The paper [18] demonstrates that for a simplified model, for example, a single-layer cylinder, the analytical solution of the three-dimensional bio-heat equation is possible with a Green's function. In our case when the object (a limb) has multiple layers, it is reasonable to use a one-dimensional model only [19–21], which helps describe the process of microwave rewarming in layers.

We will study a phantom representing a one-dimensional model consisting of five plane-parallel uniform layers with biophysical properties of human tissues (**Figure 4**). The model is idealized and enables the modification of physical properties, layer thickness and layer number. It is assumed that a plane wave is incident on the phantom surface, any loss occurs only within the volume of the warmed layer, and all the delivered power is absorbed only within the volume of the phantom.

Furthermore, we will contemplate the solution of the heat-transfer Eq. (2) without accounting for the energy of metabolism and heat transfer by the blood. This approach is justified because during hypothermia, the blood flow in tissues virtually stops and the metabolism is also minimal. Without the aforementioned terms, the Pennes equation turns into the heat-transfer equation for a solid object inside which there are heat sources, i.e. in our case, absorbed power of the EMF:

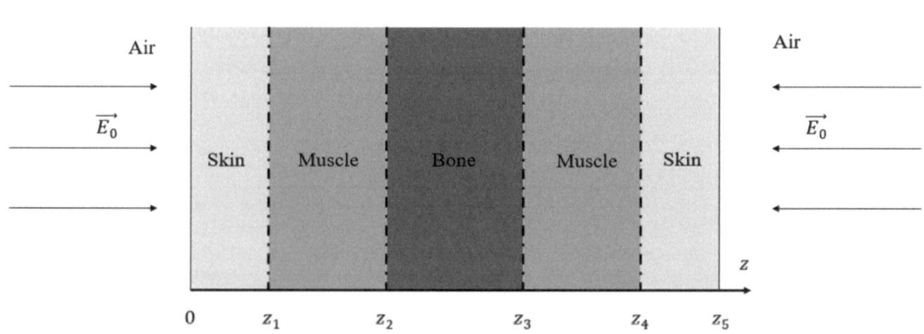

Figure 4.
A schematic diagram of the single-layer problem in question: A five-layer model of the phantom consisting of the skin, muscles, bones, muscles, skin.

$$\rho C\frac{\partial T}{\partial t} = \lambda \Delta T\left(\vec{r},t\right) + q_{MW}\left(\vec{r},t\right). \quad (3)$$

The EMF volume power for a one-dimensional problem can be written as:

$$q_{MW}(r) = \frac{1}{2}\sigma * E(r)^2. \quad (4)$$

Electric field intensity E included in the volume power expression can be determined by solving the Eq. (1). For a simplified one-dimensional model, the field can be described as the following expression:

$$E(z) = E_0 e^{-\frac{z}{d}} + E_0 e^{\frac{z-Z}{d}}, \quad (5)$$

where Z is the thickness of the rewarmed layer, and d – the thickness of the penetration depth. Because, the distribution occurs in the multi-layer environment, the value of penetration depth and distribution of the field intensity are calculated individually for each layer.

First, the initial condition defining the temperature of an entire phantom is set:

$$t = 0: \quad T(z,0) = T_0, \ \ 0 \le z \le Z. \quad (6)$$

In the phantom, at interfaces between the biological layers, the interface boundary condition equation is used and expressed as follows [22, 23]:

$$\begin{cases} -\lambda_1 \dfrac{\partial T_1(z_0,t)}{\partial z} = -\lambda_2 \dfrac{\partial T_2(z_0,t)}{\partial z} \\ T_1(z_0,t) = T_2(z_0,t) \end{cases}. \quad (7)$$

Two problems are interesting to explore. The first is to consider gravity-type airflow between the environment and the warmed object. To solve it, the convection boundary condition is used [22, 23]:

$$z = 0: \quad -\lambda \frac{\partial T(0,t)}{\partial z} = \kappa(T_{air} - T(0,t)), \ t > 0; \quad (8)$$

$$z = Z: \quad -\lambda \frac{\partial T(Z,t)}{\partial z} = \kappa(T(Z,t) - T_{air}), \ t > 0. \quad (9)$$

The second problem is related to a situation when the rewarmed object is surrounded by the medium sustained at a constant temperature, for example, by air blowing around at a specific temperature. To solve it, the specified boundary condition is used [22, 23]:

$$z = 0: \quad T(0,t) = T_{air}, \ t > 0; \quad (10)$$

$$z = Z: \quad T(Z,t) = T_{air}, \ t > 0. \quad (11)$$

In the expressions above, κ is the heat-transfer ratio that describes heat flow from one medium to the other. For natural convection air, it ranges from 1 to 25 $W/m^2{}^\circ C$ [22]. For the purpose of further calculation experiments, the value of κ = 25 $W/m^2{}^\circ C$ is chosen.

DOI: http://dx.doi.org/10.5772/intechopen.100216

	EMW frequency, MHz	Density ρ, kg/m^3	Heat capacity C, J/kg°C	Thermal conductivity λ, W/kg°C	Dielectric permittivity ε	Conductivity S/m	Penetration depth δ, cm
Skin tissue	433	1100	3150	0.25	51.3	0.72	2.9
	915				47.8	0.92	1.7
	2450				44.6	1.72	0.78
Muscle tissue	433	1050	3500	0.49	63.6	0.98	2.4
	915				60.7	1.19	1.5
	2450				55.8	2.33	0.67
Bone tissue	433	1990	2238	0.36	13.3	0.116	7.1
	915				12.5	0.18	3.9
	2450				11.7	0.41	1.6

Table 1.
Biophysical properties of human tissues.

The heat-transfer equation was numerically solved with the sweep method, which is often used to solve heat-transfer equations [24, 25]. It is one of the methods of a sequential elimination of unknowns. For the sweep method Eq. (3), partial differentials need to be approximated with corresponding finite differences, which results in the system of linear algebraic equations. To solve the heat-transfer problem in a multilayered one-dimensional phantom imitating human tissues, Mathcad 15 package was used [26] for this method.

Parameters of biological tissues that were applied to model the phantom are presented in **Table 1** [27]. The initial temperature of the cool phantom was 10°C, the surrounding air was 25°C.

The spatial parameters in the examples below are as follows: the overall phantom is 80 mm thick, the skin layer is 4 mm, the muscle layer is 28 mm, and the bone layer is 16 mm.

Figure 5 shows temperature distribution across the phantom when the phantom is rewarmed (a) without the electromagnetic field only owing to the thermal exchange with stationary air, and (b) by forced air of +25°C. The latter was proposed in the paper [28] to cool the skin layer during microwave rewarming to minimize the risk of overheating critical zones.

With microwave rewarming, the distributions change, that being considerably dependent on electromagnetic field frequencies. Corresponding thermal field distributions are shown in **Figure 6**.

The role of microwave rewarming and the dependency of the end result with the frequency value is clearer in **Figure 7**, where temperature distributions after 30-minute rewarming in the absence of the electromagnetic field and with fields of different frequencies are shown.

Critical zones, such as fingers, toes, palms, that is, a less thick phantom, demonstrate a drastically different picture. **Figure 8** shows field distributions in a 5-layer phantom with a total thickness of 2 cm after 30-minute rewarming.

Rapid rewarming is observed throughout the entire depth of the phantom and without forced moving air, there is a high risk of overheating, particularly, with the low-frequency microwave exposure.

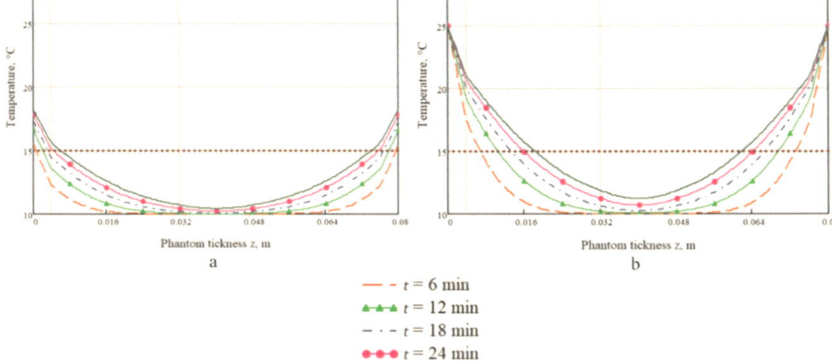

Figure 5.
Distributions of temperature fields in a one-dimensional phantom with microwave-free rewarming in stationary air (a) and by forced air (b) of 25°C.

DOI: http://dx.doi.org/10.5772/intechopen.1002216

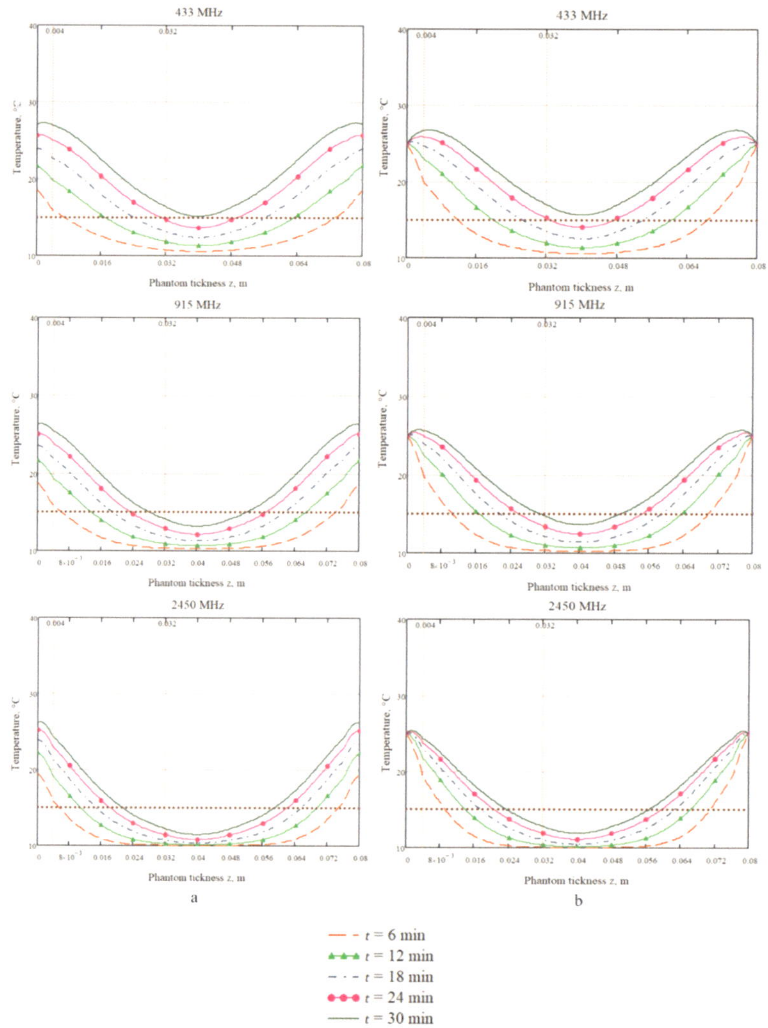

Figure 6.
Thermal field distribution in the phantom volume rewarmed in the microwave field of different frequencies (433 MHz, 915 MHz, 2.45 GHz), without forced air (a) and with forced air (b), t_{max} *= 30 min.*

4. Modeling of the chamber with a dielectric phantom inside in CST studio

As noted above, because a human limb does not lend itself easily to a mathematical description, initially, the modeling was performed by imitating it with a dielectric phantom of a more basic shape with similar dielectric permittivity and thermal conductivity. For example, the paper [29] explores field distributions in a cylindrical dielectric phantom inside a microwave chamber (**Figure 9**).

It was assumed that microwaves would be excited by a spike antenna located in the upper-right corner of the chamber side wall that is closer to the reader. **Figure 10** shows an example of the distribution of electric field amplitudes in the middle section of the chamber and phantom, a two-layer cylinder whose dielectric permittivity of the inner volume resembles the one of the muscle tissue and the outer layer is similar to the human skin and fat tissue.

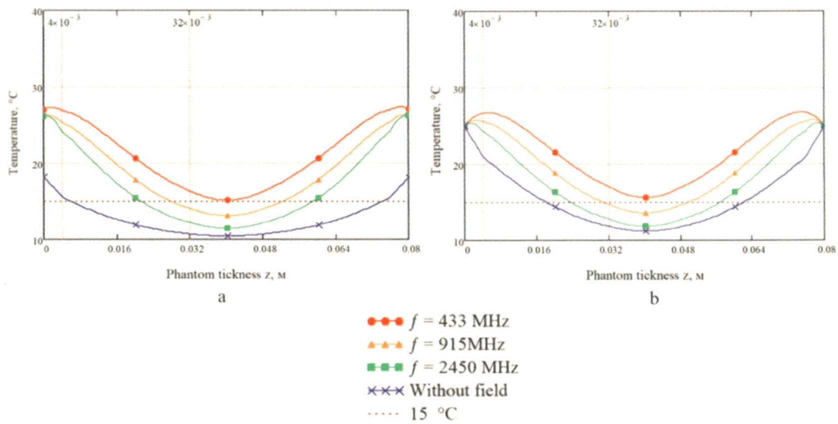

Figure 7.
Temperature distributions after 30-minute rewarming with (a) stationary and (b) forced air without the microwave field and with fields of different frequencies.

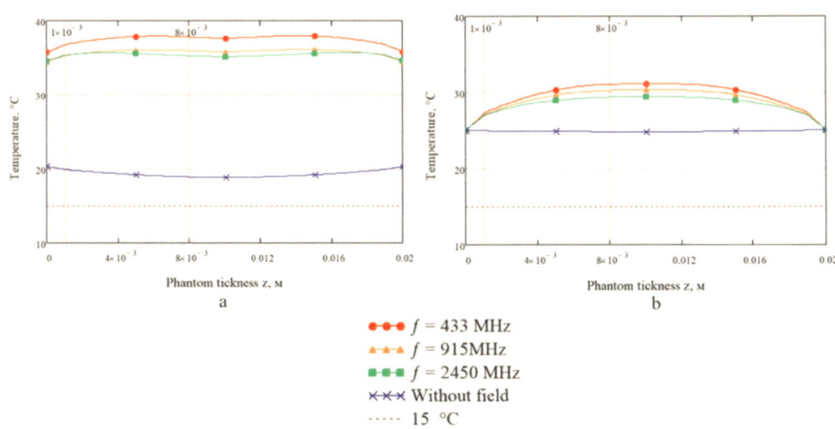

Figure 8.
Temperature distributions in a multi-layered 2-cm thick phantom after 30-minute rewarming with (a) stationary and (b) forced air with microwave fields of different frequencies.

Figure 10a clearly demonstrates that electromagnetic field in the chamber penetrates the phantom volume but its amplitude rapidly decreases. High amplitude on the right in **Figure 10b** is due to the placement of the radiator on this side of the chamber.

As noted, the VSWR, the ratio of reflection off the chamber with the phantom, contributes to the effect. **Figure 11** shows estimated VSWR dependence on the frequency in the case of a single-layer (muscle tissue) and two-layer (muscle and skin/fat tissues) phantom in the chamber. The skin and fat tissue plays the role of an "aligning" layer and decreases the VSWR value to some extent (**Figure 11**).

It is clear that around 2.45 GHz, the standing wave ratio for a two-layer cylinder is approximately 2–2.5.

5. Physical modeling (dielectric phantoms)

Practically at the same time as we were trying to describe this process with mathematical models, attempts were being made to perform physical modeling of the

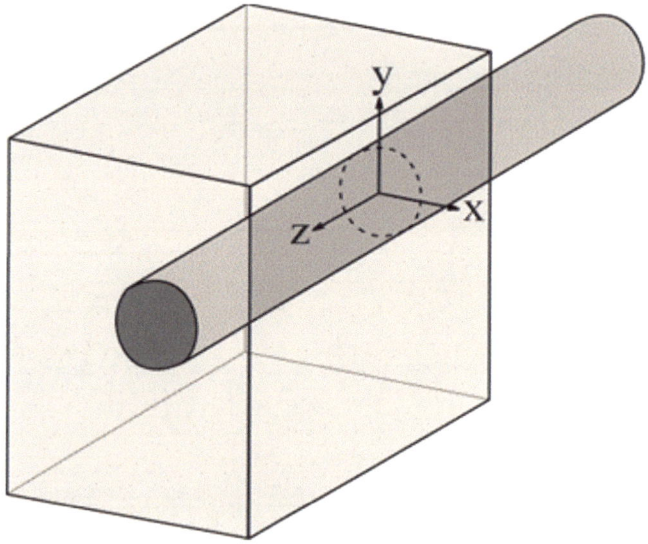

Figure 9.
A rectangular conductive chamber and a partially inserted cylindrical dielectric phantom in the protective shield—A mathematical model of a microwave chamber with a limb phantom.

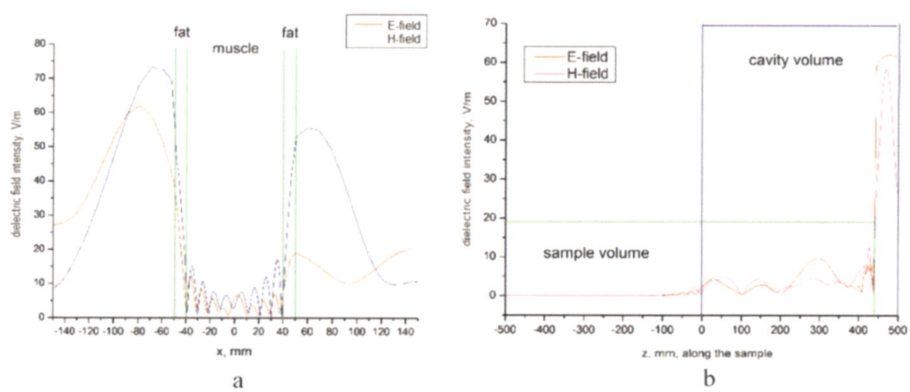

Figure 10.
Distribution of the electric field in the plane corresponding to the chamber's central section (a) along the oX axis and (b) along the central oZ axis [29].

frostbite rewarming process in dielectric phantoms imitating a human limb. Initially, it was a thin wall polymer cylinder filled with normal saline [29]. Afterwards, arm [30, 31] and leg [32] phantoms were created (**Figure 12**).

Normal saline, ground pork meat, HEC gel [33] were used as a phantom filling materials. Dielectric permittivity and conductivity of the last-named were selected to be similar to ones of the muscle tissue.

To study the distribution of temperature in the phantom volume, an experimental unit was set up, with the key components being a vertical microwave chamber (**Figure 13**) and the mechanism above that controlled a dip temperature sensor.

The phantom under investigation was placed into the chamber vertically and initial temperature distribution along the phantom's central axis was measured with the dip temperature sensor. Then the sensor was removed, the chamber (safety

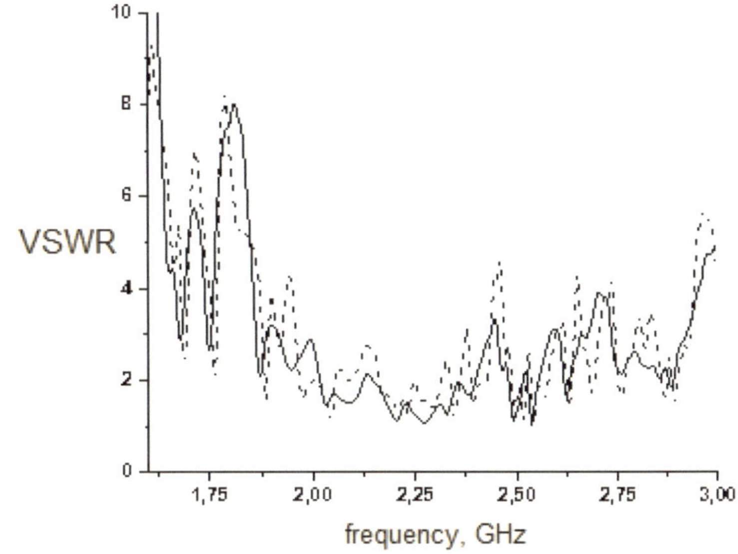

Figure 11.
The frequency dependence of the voltage standing wave ratio, the dotted line corresponds to a single-layer phantom, the solid line to a two-layer phantom [29].

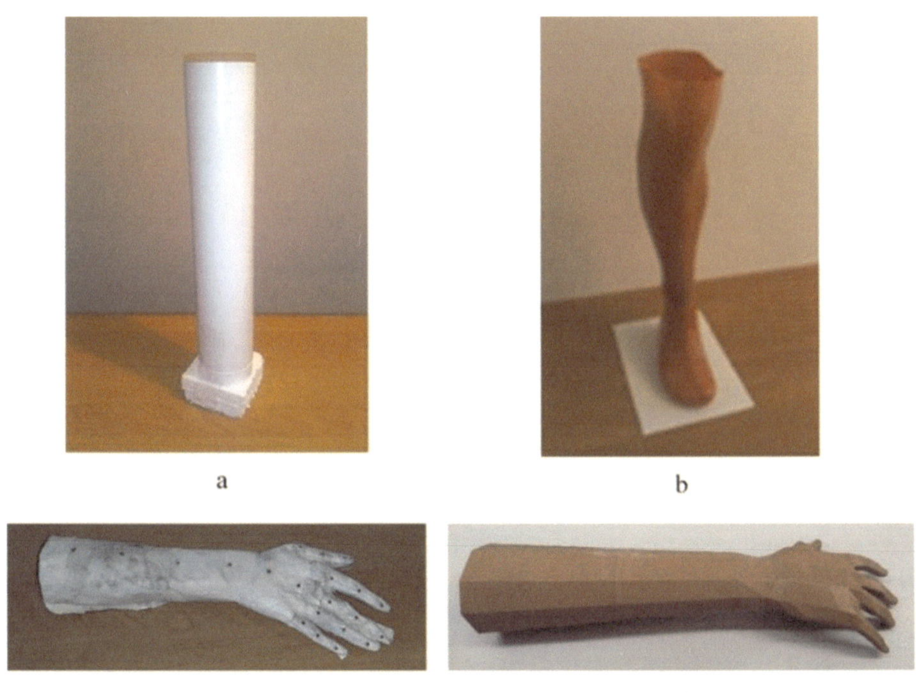

Figure 12.
Phantom shells used in the experiments.

sleeve) was closed, and microwave rewarming was initiated. Following a procedure of microwave rewarming, the sleeve was opened and the dip sensor measured temperature distribution along the phantom axis again. **Figure 14** shows measurement results

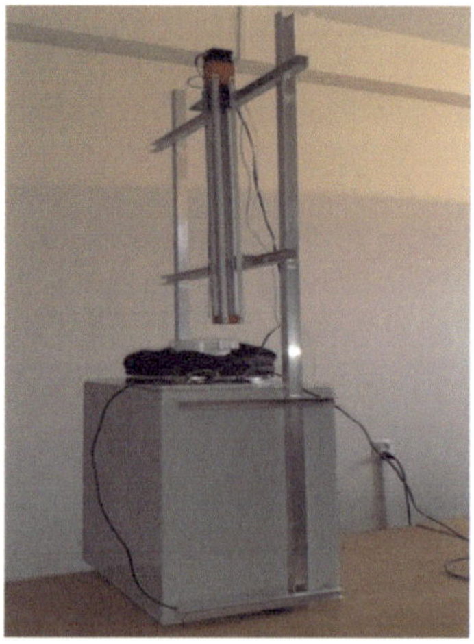

a

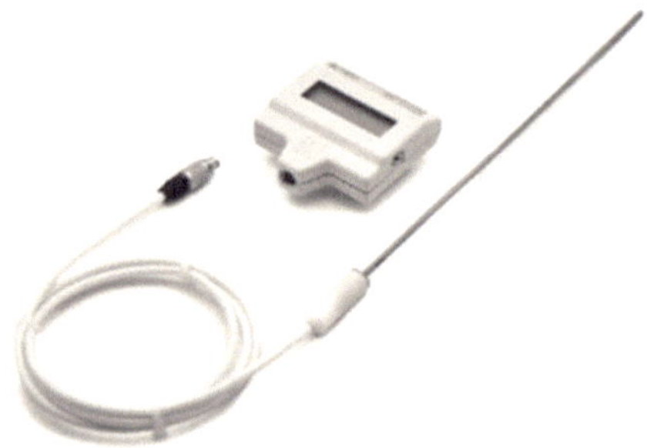

b

Figure 13.
The experimental unit to measure longitudinal temperature distributions in the volume of dielectric phantoms (a) and LT-300 dip temperature sensor (b) [30].

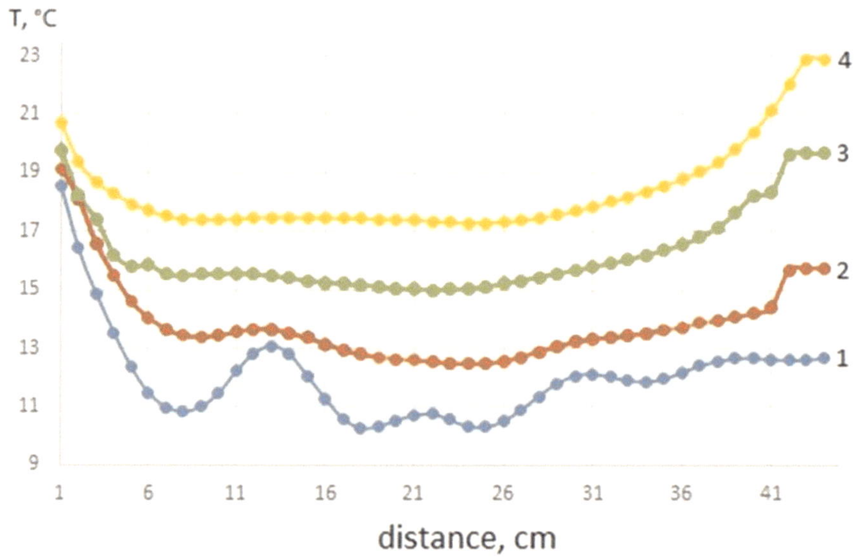

Figure 14.
Temperature distribution along the longitudinal axis of the phantom warmed up in the microwave chamber: 1 – Without microwave rewarming in the phantom center; 2,3,4 – In the phantom center after 6-minute microwave rewarming; 5 – Along the phantom wall after 6-minute microwave rewarming.

for temperature distribution in the volume of a cylindric phantom (**Figure 12a**) filled with ground pork meat. For this filler, the measured real part of dielectric permittivity was 50, imaginary part–16.9, being close to the parameters of blood-filled human muscle tissue [18].

Importantly, the distribution observed showed no considerable non-uniformities of the temperature such as "standing wave" along the central longitudinal axis and along the phantom layer near the wall. The initial non-uniform temperature distribution along the longitudinal axis (curve 1) due to non-uniform cooling of the phantom was smoothened out as the rewarming progressed. A considerable rise in the temperature was observed on the right in the zone immediately near the radiator (located in the plane that corresponds to 55 cm, according to the scale in **Figure 10**). Rapid warming observed on the left side of the figure corresponding to the open upper part of the phantom is due to the rapid thermal exchange of the filler with air.

From **Figure 14**, it is clear that after 6-minute rewarming at the generator's power of about 40 W, the temperature rise in the volume is approximately 2°C. More rapid rewarming of the filler along the phantom wall matches model expectations shown in **Figure 6**.

To reduce the intensity of the electromagnetic field near the chamber's side wall where fingers/toes are located, it was proposed to use a linear antenna a little shifted to the input aperture [30]. **Figure 15** shows a sequential process of rewarming a cylindrical phantom for such radiator location. The initial non-uniformity of the temperature along the axis of the cooled phantom is also substantial but it is smoothed out as it is warmed up. The temperature in the right zone where toes/fingers will be located does not rise as rapidly.

In **Figure 16**, measurement results are given for temperature distribution in the volume of a leg phantom (along the longitudinal axis up to the heel, excluding the foot and toes). **Figure 17** shows temperature distribution along an arm phantom, excluding fingers.

Both figures demonstrate a considerably steeper increase in the temperature in the areas of phantom narrowing (around the ankle and wrist).

The change in the temperature distribution in fingers/toes was investigated using a papier-mâché phantom (**Figure 12c**), This phantom was also filled with ground pork meat. The phantom was first cooled, with temperature distribution measurements performed at all test points (**Figure 18**), followed by three rewarming procedures, each 6 minutes long, with generator power 40 W. Temperature measurements were performed in the phantom volume with a temperature sensor LT-300 through openings in each of the marked points. Thermal field distribution was evaluated both in the upper arm and the hand, palm (dorsal side) and fingers (**Figure 19**).

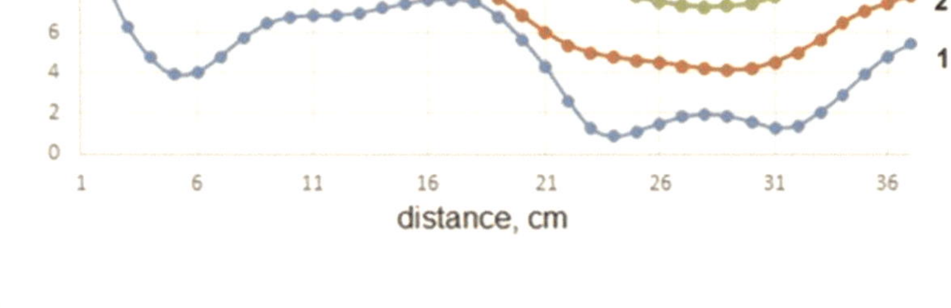

Figure 15.
Temperature distribution in the volume of a cylindrical phantom in a chamber with a linear radiator: 1 – Before microwave rewarming, 2–4 – After 3 rewarming sessions, each 6 minutes long, 40 W.

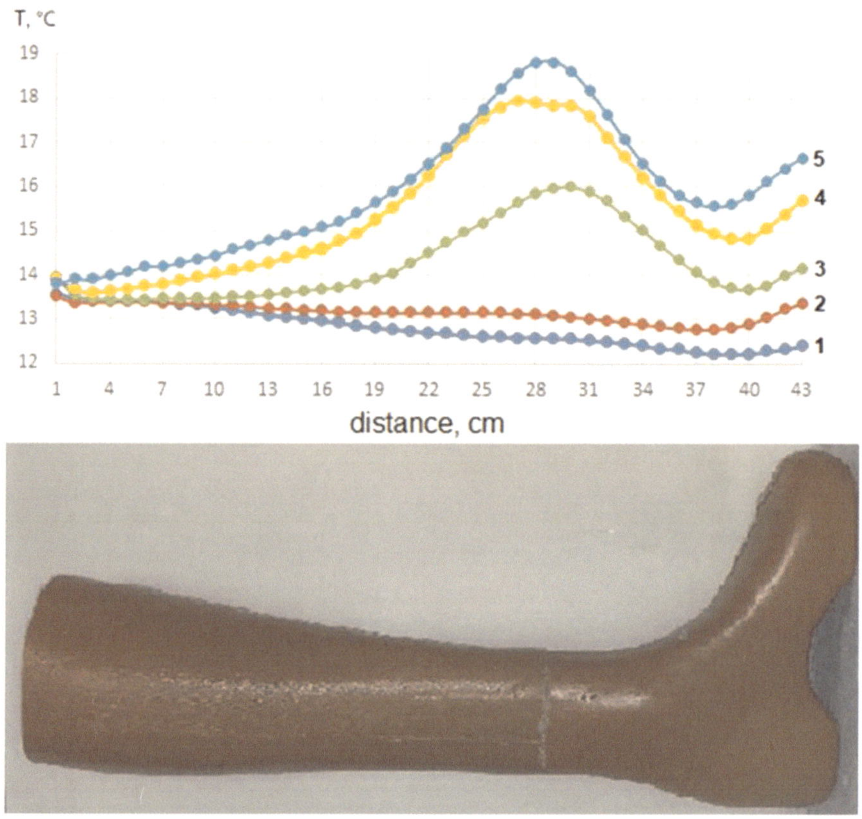

Figure 16.
Temperature measurement results along the gastrocnemius muscle along the longitudinal axis of the leg phantom: 1 – Before microwave rewarming, 2–5 – After 4 rewarming sessions, each 6 minutes long, 40 W.

First, notably, due to a smaller volume, more uniform initial cooling was achieved. The initial temperature rise in the finger area in the absence of electromagnetic field rewarming is elaborated in Part 1 of the paper as caused by thermal exchange with the environment at room temperature.

Under microwave exposure, we observe more rapid rewarming of the narrowing forearm zone (points 20, 19, 18), which was well consistent with the calculation results for the model presented above in Part 2. The temperature then rises in the critical palm zone (points 17, 16) and proximal and middle finger phalanges (points 3, 2, 6, 5, 9, 8, 12, 11, 15, 14) and decreases in distal phalanges (points 1, 4, 7, 10, 13). The decrease is caused by two factors: smaller electromagnetic intensity in this chamber area due to the location of the radiator more closely to the opening (see **Figure 15** for distribution in the cylindric phantom) and more rapid thermal exchange between the fingers and air, as noted above.

6. Some findings of experiments with volunteers

Approval of the Tomsk State University Committee of Bioethics was obtained to involve informed volunteers in this study, considering widespread, years long practice

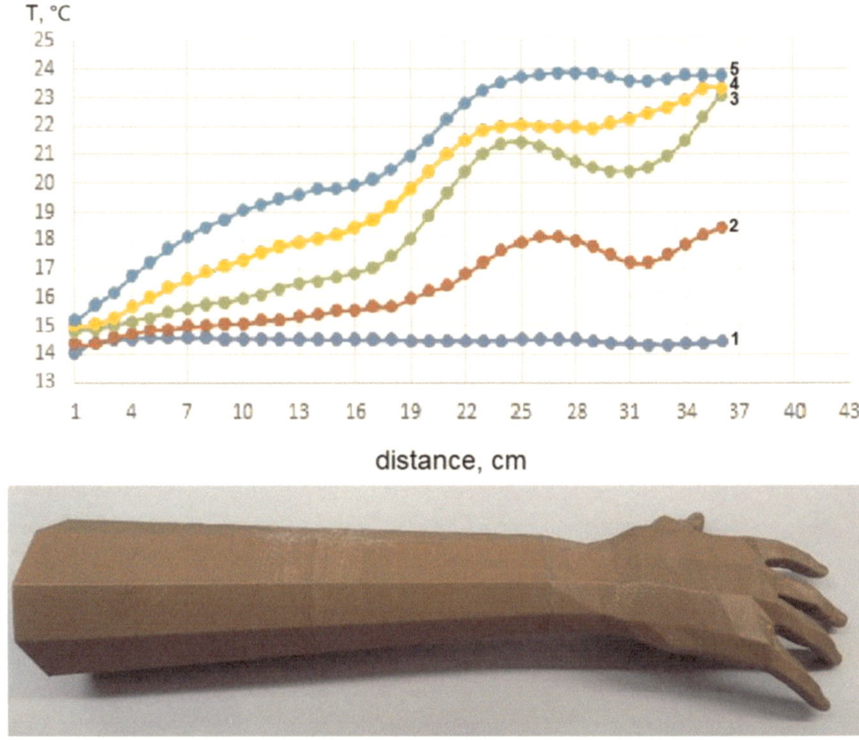

Figure 17.
Temperature measurement results in the volume of the arm phantom: 1 – Before microwave rewarming, 2–5 – After 4 rewarming sessions, each 6 minutes long, 40 W.

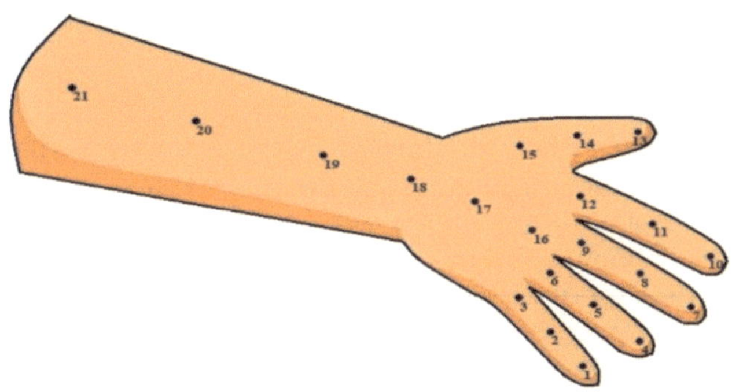

Figure 18.
A view of the arm phantom with indicated temperature measurement points.

of microwave applications in physical therapy and hyperthermia and positive effects of frostbite microwave treatment of animal limbs [15] and demonstration of the method in phantoms. The volunteers invited to participate in the study were emergency patients admitted to Tomsk Medical Sanitary Unit No. 2 and Tomsk Municipal Hospital No. 3.

The treatment technique included thermal insulation of the cold-injured area prior to microwave rewarming, followed by 1–3 30-minute rewarming procedures in the microwave chamber. For this purpose, a generating unit of a SMVI-200 medical

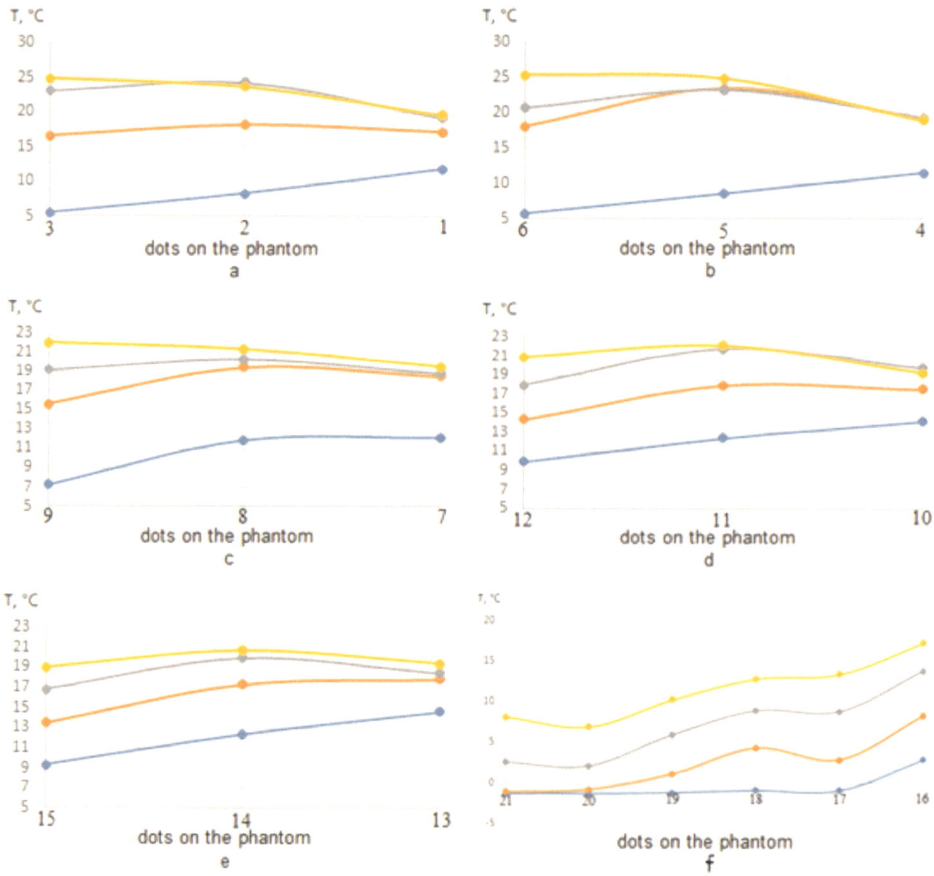

Figure 19.
Temperature distribution throughout an arm phantom: Pinky finger (a), ring finger (b), middle finger (c), index finger (d), thumb (e), forearm (points 21, 20, 19), wrist (point 18) and palm (points 17, 16) (f).

physiotherapeutic device was used with generator power of 30–60 W. Routine drug treatment for such injuries was administered.

In 2019–2022, a total of 14 patients underwent treatment with this technique [34, 35]. For the majority of patients, the technique indeed helped prevent amputation altogether or considerably lessen the scale of it. The compelling demonstration of the technique's efficacy is the case of a patient [35] who, after being outside in the cold for many hours, was indicated to have both hands (**Figure 20a**) and both feet (**Figure 20b**) amputated.

Following the microwave treatment, the patient was discharged only with the toe distal phalanges amputated (**Figure 21**).

7. Conclusion

As a result of mathematical modeling of the microwave rewarming in a chamber and experiments in dielectric phantoms, appropriate modes were identified to ensure safe and effective deep rewarming of a frostbitten limb. The efficacy of the technique was demonstrated in informed volunteers. The treatment consisted in the microwave

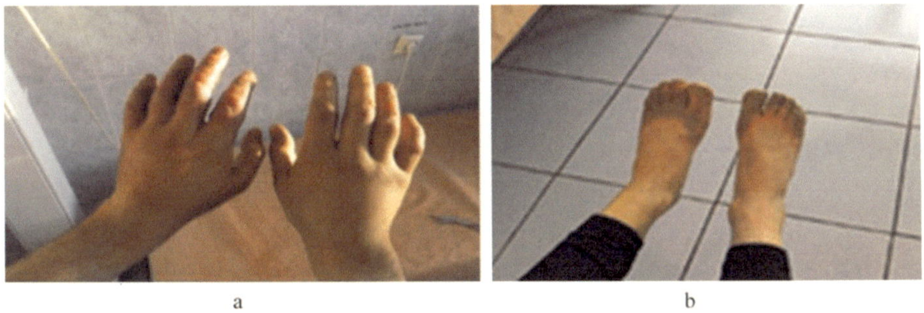

Figure 20.
Cold-injured limbs: (a) hands and fingers, (b) feet and toes [35].

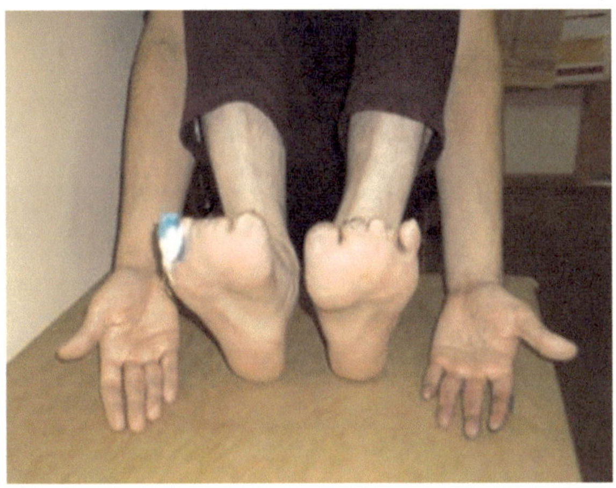

Figure 21.
Outcomes of microwave rewarming treatment: Amputation scale is minimized, both hands and feet are salvaged, only toe distal phalanges are amputated [35].

frequency of 2.45 GHz permitted for medical use and power that is less than the one used in physiotherapy procedures. The most efficient procedure was the combination of microwave rewarming and a surgical procedure (fasciotomy, onychectomy), as well as haemorheologic drugs and vasodilators. It is of outmost importance to use this approach in the early reactive period and have no skin rewarming of the affected limb at all prior to microwave rewarming. It is also recommended to perform thermal insulation limb before microwave rewarming.

Clearly, both the chamber and the microwave generator should be optimized, which requires solution of a number of other electrodynamic, thermophysical, and engineering problems. However, the available results already demonstrate a novel, useful, and noteworthy application of microwave radiation in medicine.

Acknowledgements

The work is supported by the Tomsk State University Development Program (Priority-2030).

DOI: http://dx.doi.org/10.5772/intechopen.1002216

Conflict of interest

The authors declare no conflict of interest.

Notes/thanks/other declarations

The authors thank Professor Gavrilin E.V., and Antipov V.B. for significant contribution, active participation and support of this work.

Author details

Grigory Dunaevskiy*, Alexander Nechaev, Polina Smygalina and Igor Dorofeev
National Research Tomsk State University, Tomsk, Russian

*Address all correspondence to: proecs@mail.tsu.ru

References

[1] Lorentzen AK, Davis C, Penninga L. Interventions for frostbite injuries. Cochrane Database of Systematic Reviews. 2020;**12**(12):CD012980. DOI: 10.1002/14651858.CD012980.pub2

[2] Shapovalov KG et al. The damage of endothelial cells and track record cytokines at patients in different periods of local frostbite. Bulletin of the East Siberian Scientific Center SBRAMS. 2006;**6**(52):126-128

[3] Kotel'nikov VP. Frostbites. Moscow: Meditsina; 1988. p. 256

[4] Gavrilin EV. Regional Disorders of Intraosseous Hemodynamics in Pathogenesis and Treatment of Extremity Cryoinjury [Thesis]. Tomsk: Author's Abstract of Doct. Medical Sci; 2001

[5] Habash RW, Bansal R, Krewski D, Alhafid HT. Thermal therapy, part 1: An introduction to thermal therapy. Critical Reviews in Biomedical Engineering. 2006;**34**(6):459-489. DOI: 10.1615/critrevbiomedeng.v34.i6.20

[6] Habash RW. Bioeffects and Therapeutic Applications of Electromagnetic Energy. Boca Raton: CRC Press; 2008. DOI: 10.1201/9781420062854.ch5

[7] Veepsa B. Antennas and Other Electromagnetic Applicators in Biology and Medicine. India: Department of Electronics & Communication Engineering Bharati Vidyapeeth's College of Engineering; 2005

[8] Balanic CA. Chapter 27. Antennas for medical therapy and diagnostics. In: Modern Antenna Handbook. USA: Wiley; 2008. pp. 1377-1419

[9] Physiotherapy Apparatus CMWi-200. Available from: https://medteco.ru/physiotherapy/centimeter-wave-therapy/smvi-200/ [Accessed: April 2, 2023]

[10] Shortwave and Microwave Diathermy. Deep tissue heating. Available from: https://www.btlnet.com/diathermy [Accessed: April 2, 2023]

[11] Falk MH, Issels RD. Hyperthermia in oncology. International Journal of Hyperthermia. 2001;**17**(1):1-18. DOI: 10.1080/02656730150201552

[12] Giombini A, Giovannini V, Di Cesare A, Pacetti P, Ichinoseki-Sekine N, Shiraishi M, et al. Hyperthermia induced by microwave diathermy in the management of muscle and tendon injuries. British Medical Bulletin. 2007;**83**:379-396. DOI: 10.1093/bmb/ldm0204

[13] Dutta J, Kundu B. Two-dimensional hybrid analytical approach for the investigation of thermal aspects in human tissue undergoing regional hyperthermia therapy. Proceedings of the Institution of Mechanical Engineers, Part C: Journal of Mechanical Engineering Science. 2020;**234**(20): 3951-3966. DOI: 10.1177/0954406220919460

[14] Antipov VB et al. Device for Treatment of Frostbites of the Extremities. R.F. Patent No. 2334494 C2 (A 61). Off. Bull. Invent. Utility Models Moscow. Bull. 27. Moscow: Rospatent, Federal Service for Intellectual Property; 2008

[15] Gorelik IÉ. Prophylaxis against Necrosis in the Frostbitten Extremities during Pre-Reactive and Early Reactive Periods: Author's Abstract of Cand. Kemerovo: Medical Sci. Diss; 2010. p. 22

[16] Antipov VB, Dunaevskiy GE, Gavrilin EV. Device for treatment of limb freezers. R.F. Patent No. 170090 U1 (UM). Off. Bull. Invent. Utility Models. Moscow. Bull. 11. Moscow: Rospatent, Federal Service for Intellectual Property; 2017

[17] Pennes HH. Analysis of tissue and arterial blood temperatures in the resting human forearm. 1948. Journal of Applied Physiology (Bethesda, MD: 1985). 1998; **85**(1):5-34. DOI: 10.1152/jappl.1998.85.1.5

[18] Vyas R, Rustgi ML. Green's function solution to the tissue bioheat equation. Medical Physics. 1992;**19**(5):1319-1324. DOI: 10.1118/1.596767

[19] Foster KR, Lozano-Nieto A, Riu PJ, Ely TS. Heating of tissues by microwaves: A model analysis. Bioelectromagnetics. 1998;**19**(7): 420-428. DOI: 10.1002/(sici)1521-186x(1998)19:7<420::aid-bem3>3.0.co;2-3

[20] Kanezaki A, Hirata A, Watanabe S, Shirai H. Parameter variation effects on temperature elevation in a steady-state, one-dimensional thermal model for millimeter wave exposure of one- and three-layer human tissue. Physics in Medicine and Biology. 2010;**55**(16): 4647-4659. DOI: 10.1088/0031-9155/55/16/003

[21] Rodrigues DB, Pereira PJ, Limão-Vieira P, Stauffer PR, Maccarini PF. Study of the one dimensional and transient bioheat transfer equation: Multi-layer solution development and applications. International Journal of Heat and Mass Transfer. 2013;**62**: 153-162. DOI: 10.1016/j.ijheatmasstransfer.2012.11.082

[22] Çengel Y, A, Ghajar AJ. Heat and Mass Transfer: Fundamentals & Applications. Fifth ed. New York City: McGraw Hill Education; 2015. p. 992

[23] Lykov AV. Theory of Thermal Conduction. Moscow: Higher School Publishing House; 1966. p. 600

[24] Kuznetsov GV, Shermet MA. Difference Methods of Solution of Heat Conduction Problems. Tomsk: TPU; 2007. p. 172

[25] Demidovich BP. Numerical Methods of Analysis. Moscow: Fizmatgiz; 1963. p. 400

[26] Maxfield B. Engineering with Mathcad: Using Mathcad to Create and Organize your Engineering Calculations. Oxford: Butterworth-Heinemann; 2006. p. 494

[27] Gabriel S, Lau RW, Gabriel C. The dielectric properties of biological tissues: II. Measurements in the frequency range 10 Hz to 20 GHz. Physics in Medicine and Biology. 1996;**41**(11):2251-2269. DOI: 10.1088/0031-9155/41/11/002

[28] Antipov VB, Dunaevskiy GE, Gavrilin EV. Device for treatment of Frostflow threats. R.F. Patent No. 188862 U1 (UM). Off. Bull. Invent. Utility Models. Moscow. Bull. 12. Moscow: Rospatent, Federal Service for Intellectual Property; 2019

[29] Antipov VB, Gavrilin EV, Dorofeev IO, et al. Distributions of electric and thermal fields in a rectangular microwave chamber with a cylindrical phantom. Russian Physics Journal. 2020;**63**:196-203. DOI: 10.1007/s11182-020-02021-7

[30] Nechaev AN, Dunaevskiy GE. Temperature distribution in the phantom of limb in the microwave chamber with linear source of radiation. Journal of Physics: Conference Series. 2021;**1989**:012035

[31] Nechaev AN, Dunaevskij GE. Distribution of temperature field in

phantom of hand with microwave radiation. Journal of Physics: Conference Series. 2021;**2140**(1):012034

[32] Dunaevskiy GE, Nechaev AN, Badin AV, Teterina DD. Distribution of temperature along the phantom of the human leg under the influence of microwave radiation. Journal of Physics: Conference Series. 2020;**1499**:012021

[33] Hartsgrove G, Kraszewski A, Surowiec A. Simulated biological materials for electromagnetic radiation absorption studies. Bioelectromagnetics. 1987;**8**(1):29-36. DOI: 10.1002/bem.2250080105

[34] Dunaevskiy G, Gavrilin E, Pomytkin A, et al. Reduction of amputations of frostbitten limbs by treatment using microwave rewarming. Scientific Reports. 2023;**13**:1362. DOI: 10.1038/s41598-023-28535-x

[35] Gavrilin EV, Dunaevskiy GE, Antipov VB. Microwave treatment of cold injuries. Journal of Emergencies, Trauma, and Shock. 2021;**14**(2):108-110. DOI: 10.4103/JETS.JETS_142_20

Chapter 3

Repeatability Assessment of the Wavelia#2 Microwave Breast Imaging Scan: Experimental Performance Analysis Prior to Clinical Investigation

Angie Fasoula, Petros Arvanitis and Luc Duchesne

Abstract

Microwave imaging is an emerging imaging modality with the potential to support the diagnosis of breast cancer. Over the last two decades, a notable number of MicroWave Breast Imaging (MWBI) prototype devices have been developed and experimentally tested in Europe, North America and Asia. A small number of prototypes are currently in large-scale clinical investigations towards the demonstration of clinical efficacy, as well as identification of the clinical cases for which MWBI could bring added value over the existing breast imaging modalities. In this chapter, the methodology employed for quantitative assessment of the Wavelia#2 MWBI system reliability based on experimental data is presented. The analysis includes an assembly of the most representative findings from the series of the MWBI experimental tests performed on breast phantoms after installation of the Wavelia#2 investigational device at Galway University Hospital (GUH) – Symptomatic Breast Unit for a Phase-II Pilot clinical investigation. To the authors' knowledge, the notion of MWBI scan and reconstructed image repeatability assessment has never been addressed before in the MWBI state-of-the-art.

Keywords: microwave imaging, breast cancer diagnosis, radar detection theory, anthropomorphic breast phantoms, scan repeatability, image similarity

1. Introduction

MicroWave Breast Imaging (MWBI) is an emerging imaging modality with potential to support the diagnosis of breast cancer [1–3]. Microwave imaging uses the scattered wave, or reflected wave, that arises from the contrast in dielectric properties between the various breast tissues, in the microwave frequency range. The increased volume of water within the denser breast tissues is responsible for the detectable electromagnetic scattering associated with microwave imaging. The increase of sodium and water, particularly in-bound water within the tumour cells, is expected to lead to even greater conductivity and permittivity of the tumourous tissues [4, 5]. Due to the dielectric contrast, back-scattered radar signals are physically generated, when the breast is illuminated with electromagnetic waves in the microwave frequency range.

During the last two decades, MWBI has been extensively investigated as a novel modality for the detection of breast disease, offering a non-ionizing, non-compressive approach [6] and as a potential diagnostic management strategy in the monitoring of neoadjuvant chemotherapy [7]. The non-invasive and non-ionizing characteristics of microwaves are associated with the important advantage of allowing for frequent scans of the breast using microwave imaging, unlike mammography [8]. In addition to safety, microwave imaging has further potential advantages, as it does not require uncomfortable breast compression [9].

Numerous research groups have developed and optimized hardware prototypes for microwave breast imaging, however, most of these prototypes remained in pre-clinical stage [10, 11]. While each of these devices has a distinct approach to generating images of the breast, the underlying microwave technology and the device characteristics are similar. Based on recent reviews [12–14], a total of no more than 10 MWBI system prototypes have been employed in human subject tests, to investigate the clinical utility of the MWBI technology. Those trials differ significantly in their scale, ranging from few volunteers to hundreds of patients. While most of the studies have been too small to determine clinical efficacy, larger-scale trials have been conducted with three state-of-the-art MWBI system prototypes: MARIA [15], Mammowave [16–18] and SAFE [19–21]. It should be noted that there is a fourth MWBI system with a remarkable history of testing on hundreds of breast cancer patients; it is the MWBI system developed at Dartmouth College, USA [7, 22], to our knowledge though, there have been no recent publications reporting clinical trial results with this system.

Despite favourable clinical results being reported, several recurrent limitations remain unresolved across even the most recently published clinical studies with MWBI, likely hindering the migration of this modality to the clinical setting. These are detailed in **Table 1**.

Shortcomings of State-of-the-art Microwave Breast Imaging
A non-negligible false positive rate.
The challenges of managing a wide range of breast sizes with the same MWBI system.
The automated and repeatable/consistent detection of breast pathologies of various types in breasts of various levels of density.
Factor analysis (breast density, breast size, age, cancer type and stage) in the absence of consistent datasets from large-scale MWBI clinical trials.
The identification of clinical cases where the addition of MWBI would demonstrate to be a useful clinical adjunct to detect or characterize breast pathology.
The detectability of small, non-palpable, breast pathologies.
The achievable accuracy of lesion localization in the breast has not been quantified with MWBI by any group.
The standardization of the patient positioning and scan process.

Table 1.
Microwave breast imaging: Shortcomings of the state-of-the-art.

In addition, despite the extensive efforts to harness the potential of the MWBI modality, spanning over 40 years when considering all the preliminary experimental pre-clinical developments of the various groups, a pertinent clinical application for this modality has yet to be identified.

The above-reported aspects justify further clinical research with alternative MWBI systems, such as Wavelia™ [23, 24]. In the First-in-Human (FiH) clinical study (NCT03475992), which has been conducted in a small cohort of 24 patients, the Wavelia#1 prototype system demonstrated the potential to detect and discriminate between palpable benign and malignant breast lumps, while the imaging procedure had no safety issues and patients reported a favourable experience of the test. The promising findings from this study, which provided initial data to support a valid clinical association for the Wavelia™ breast imaging technology, have warranted the preparation of further clinical investigations with an upgraded prototype version of the Wavelia™ system (Wavelia#2). The clinical data that will be further collected with Wavelia#2 is intended to build upon the outcomes of the FiH study of Wavelia#1 in larger and more diverse patient populations, as well as further address the current limitations of the state-of-the-art MWBI technology applied to clinical trials by now, as reported in **Table 1**. In this chapter, an evaluation of the Wavelia#2 MWBI system performance based on realistically complex experimental anthropomorphic breast phantoms is presented.

Even though several breast phantom studies, performed in a purely experimental setup within engineering laboratories, have been published so far [25–27], no systematic experimental evaluation performed on a MWBI prototype after installation for testing in the clinical setting, has ever been published before, to our knowledge. Among the three above-referenced MWBI prototypes, which have been employed in large-scale clinical investigations in the recent years, an experimental evaluation study has been very recently published only for the Mammowave system [28]. In this study, a simplistic phantom structure and experimental setup have been implemented though. A cylindrical plastic container filled with a low-permittivity liquid was used to mimic the adipose tissue of the breast. A glass tube, filled with a very high-permittivity liquid, was inserted in the plastic cylinder to mimic a tumourous inclusion. The whole structure was held in the prototype using a cylindrical holder, which is different from the cup that is used to hold the real breast in the clinical setting.

To our knowledge, realistically anthropomorphic, MRI-derived, 3D-printed breast phantoms [29], heterogeneously filled with adipose and fibro-glandular tissue-mimicking liquids as per [30] and including malignant and benign solid tumour phantoms with breast tumour-like irregular shapes as per [25, 31] and a realistic permittivity profile, have only been employed for the Wavelia™ system [32] so far. A series of achieved tumour detectability and imaging results on tests performed in Galway University Hospital after installation of the Wavelia#1 prototype for its FiH study were earlier published in [33].

The methodology and metrics employed for the evaluation of the Wavelia#2 MWBI system reliability are presented in the sequel of this chapter. The demonstration includes an assembly of some of the most representative findings from the series of imaging tests that were performed in a systematic manner, on the same complex anthropomorphic breast phantoms as earlier developed for Wavelia#1 [32], after installation of the Wavelia#2 investigational device at Galway University Hospital (GUH) – Symptomatic Breast Unit for a clinical investigation. To this end, the achievable level of repeatability of the Wavelia#2 scan and reported imaging findings is a key element and has thus been prioritized in the experimental evaluation of the system, towards acceptance at the clinical investigation site.

To the authors' knowledge, this is the first time that the notion of MWBI scan repeatability/reliability is ever being conceived and quantitatively assessed. A quite extensive review on the qualitative and quantitative assessment methods earlier employed on MWBI system prototypes has been recently published by Porter et al. [34] and confirms the previous statement.

2. Wavelia™ microwave breast imaging

2.1 The principles

The Wavelia#2 MWBI system operates using an array of 21 wideband probes in a horizontal circular configuration. The probes are located outside of a cylindrical container that hosts a creamy liquid, serving as a transition medium between the scanner and the breast. The liquid has dielectric properties akin to the ones of the human skin, such that it favours the penetration of the Electro-Magnetic (EM) waves in the breast. With the patient lying in a prone position, one breast is submerged at a time into the cylinder, which is filled with the liquid. The probe array is piloted to perform a vertical motion, such that the full vertical extent of the breast is included in a cylindrical scan of the imaging scene. At each 2 mm vertical scan position, each antenna illuminates individually the breast using EM waves in the frequency range [0.8–4] GHz, while the remaining antennas receive the electromagnetic scattering at various angles around the circle in a multi-static radar imaging system configuration. This data is used to form a 3D image of the dielectrically contrasted interior breast tissues.

As earlier introduced in [33, 35], due to the lossy profile of both the breast tissues and the employed transition medium, but also due to the distinct physical meaning of the forward and backward scattering mechanisms, sectorization of the imaging scene is performed in Wavelia™, to more efficiently illuminate and reconstruct the imaging scene. A subset of 8 out of the 21 probes of the Wavelia#2 system is used each time to numerically form a sub-array, the full multi-static data matrix of which is used to form a partial image of the breast per azimuthal sector of illumination. The process is repeated 21 times; at each repetition, the centre of the sub-array is numerically moved to the next, adjacent, probe. This way, a total of 21 partially overlapping sub-images of the breast are formed and subsequently stitched to reconstruct a coronal slice of the pendulous breast, at each vertical position of the probe array.

Wavelia™ employs the Time-Reversal MUltiple SIgnal Classification (TR-MUSIC) algorithm [36] to form the aforementioned partial images ("sub-images") of the breast. For the formation of a single TR-MUSIC sub-image of the breast at each sector and each frequency, the noise eigenvectors of the Time-Reversal Operator (TRO) are projected on the illumination vector of the sectorized probe sub-array. The scalar electric field approximation is employed, considering the co-polar component of the electric field, as radiated by the antennas of the system, to be dominant. The Hankel function of 1st kind and 0th order is then used as a basic propagation model to define each element of the illumination vector on each pixel of the imaging scene that lies within the transmitting (Tx) sensor fidelity zone, as introduced in [37].

The breast tissues are associated with high levels of power loss of the propagating electromagnetic waves, in the microwave frequency range. This is why, prior to radar imaging of the interior of the breast, pre-processing of the backscattered signals is required, to remove artefacts and unwanted interference signals, and enhance the useful radar echoes of weak-power level, originating from the interior of the breast.

DOI: http://dx.doi.org/10.5772/intechopen.1002506

The strong received signals mainly consist of direct coupling between the antennas, skin reflections, and antenna reverberation.

The speed of propagation of electromagnetic waves through the breast tissues is determined by the dielectric properties (permittivity) of these tissues. In turn, the wave speed determines the time delay for reception of the waves, which were scattered by the tissues. The time delay (or phase shift of the EM waves) is mapped to a physical distance between the transmitting/receiving antennas and the scatterer. The distance information drives the image formation, i.e. spatial mapping of any scattered radar echoes in the patient's breast. A large variability exists in the dielectric properties of each breast tissue type, over the population, as demonstrated by multiple studies involving ex-vivo dielectric measurements of a large sample of excised breast tissues [38, 39]. Considering the lack of a priori information on the dielectric consistency of each patient's breast (full dielectric map of the breast), data-driven techniques are employed in Wavelia™ to deduce the speed of the EM waves for MWBI imaging (a parameter called pc_fib is employed for image formation, as introduced in [23, 35]).

2.2 The second generation prototype

The second generation prototype of the Wavelia™ MWBI system (Wavelia#2), depicted in **Figure 1**, is a significantly upgraded version of the first generation prototype (Wavelia#1), in terms of enhanced measurement stability, better control on the patient positioning, potential to reliably scan larger volumes of breasts and an improved ergonomic design.

The technical upgrade of the Wavelia™ prototype has been designed based on the experience gained from the FiH clinical investigation. It is intended to ensure enhanced performance, mainly in terms of better repeatability of the MWBI scan, and thus a higher confidence level for the information content of the reconstructed breast images and their diagnostic value. Enhanced sensitivity to weaker signals, to ultimately support detection of smaller lesions and a better accuracy of the lesion localization is also targeted with the Wavelia™ technical upgrades.

The following are the main adaptations which have been implemented in the transition from the Wavelia#1 to the Wavelia#2 prototype:

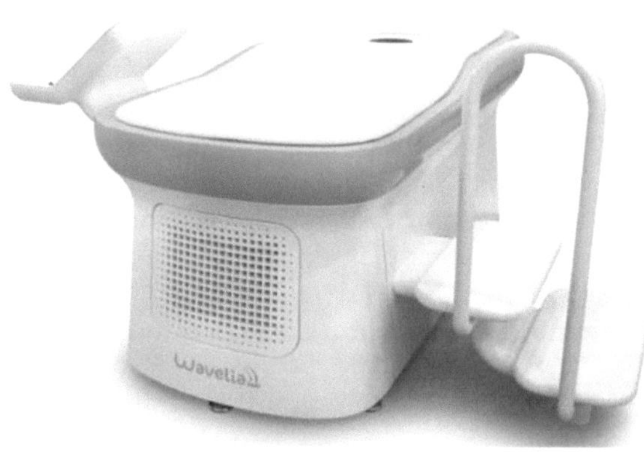

Figure 1.
The Wavelia#2 microwave breast imaging prototype.

I. Adjustment in the size of the container of the scanner to facilitate better scan quality for larger breasts. To be noted though that breast volumes larger than >1000 mL have been scanned with Wavelia#1 during the FiH study already. An upper size limit of ~500 ml has been earlier reported for the MARIA M5 prototype [15]. No information on the breast size limits of the subsequent prototypes of MARIA (M6 and M7), nor of the Mammowave and the SAFE systems, have been more recently published. The range of breast volumes and thus the portion of the women population for whom MWBI is appropriate with each implementation of the technology remains yet open to be specified.

II. Design and integration of smaller antennas to improve imaging of the posterior part of the breast, i.e., reduce the distance between the uppermost scanning position and the examination table.

III. Use of upgraded Radio-Frequency (RF) components, cabling, and mechanical support of the probe array.

IV. Chemical stabilization of the transition liquid, as well as automation of its injection and suction from the cylindrical container of the scanner.

V. Better control on the environmental conditions, in the examination room and within the scanner itself, during the breast scan.

VI. Better control of the breast positioning in the scanner.

Wavelia#2 is currently being tested in a Phase-II pilot clinical investigation, at Galway University Hospital, Ireland, targeting to include approximately 70 patients. (NCT05757427).

2.3 The Wavelia™ breast imaging methodology revisited

The principles of the multi-static radar imaging methodology that is employed for the generation of the Wavelia#2 MWBI images are the same as earlier conceived for Wavelia#1 [35, 40], yet adapted to the Wavelia#2 geometry (i.e. an increase of the internal radius of the system by 11 mm, combined with an increase of the number of probes from 18 to 21). The Wavelia#2 data processing flow is revisited in **Figure 2**. The blocks in which methodological evolutions have been recently integrated for Wavelia#2, are highlighted in the block diagram and discussed in the sequel of this section. Only the MWBI image formation part of the Wavelia™ breast imaging methodology and not the image post-processing for automated Region-Of-Interest (ROI) detection [40], is part of the experimental analysis on the breast phantom MWBI scans, presented in the following sections.

As highlighted in **Figure 2**, the key evolutions that have been performed as part of the adaptation of the data processing flow for Wavelia#2 are the following:

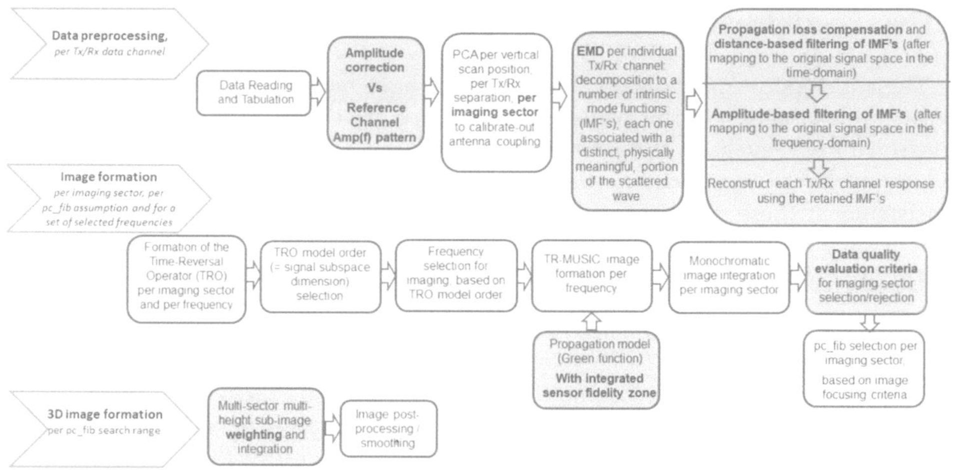

Figure 2.
Wavelia#2 data processing flow: Blocks evolved since Wavelia#1 are highlighted.

I. The *normalization of the amplitude pattern of each Tx/Rx channel with respect to a reference channel*, thus eliminating the impact of the evolution of the conducted RF paths over the frequency band.

II. The *estimation of the antenna coupling* as the first principal component, while performing *Principal Component Analysis (PCA), per imaging sector*. In Wavelia#1, PCA was performed on data from the full probe array having a fixed distance between Transmitter and Receiver (Tx/Rx separation) [35].

III. The employment of the *Empirical Model Decomposition (EMD) method* [41] to define the signal components to be further filtered or retained based on distance and/or amplitude pattern criteria.

IV. The *integration of the notion of the sensor fidelity zone [37] in the TR-MUSIC imaging algorithm* implementation [35, 36] for the Wavelia#2 prototype. The sensor fidelity zone is specified as a binary indicator of the pixels in the imaging scene, which are compliant with two constraints: (i) the propagation loss on the pixel does not exceed a certain Maximum Attenuation Level (MAL), (ii) the co-polar component of the radiated electric field dominates the other two field components at least by a ratio Minimum Polarization Ratio (MPR), such that the scalar field approximation can be considered relevant for the TR-MUSIC algorithm. The sensor fidelity zone has been re-defined for Wavelia#2 based on simulations of the radiated EM field in the CST simulation environment.

In addition to the aforementioned methodological evolutions, robustification, numerical stabilization, as well as significant acceleration of the computation have been achieved as part of the adaptation of the MWBI imaging algorithm to the Wavelia#2 prototype. This upgraded implementation is a step ahead towards standardization of the Wavelia™ breast images and more efficient processing of the data during the ongoing clinical investigation, as compared to the FiH study.

3. The experimental validation campaign

The anthropomorphic breast phantoms that have been used for the experimental testing and validation of the Wavelia#2 prototype are the same as the ones earlier developed for Wavelia#1 and presented in detail in [32, 33]. Despite the limitations and complexities introduced to the repeatability analysis by the hybrid (liquid/solid) consistency of these phantoms, to our knowledge, these phantoms represent the current state-of-the-art for experimental Microwave Breast Imaging.

The Wavelia#2 MWBI scan repeatability analysis has been structured as follows:

I. *Employ two breast phantoms, Breast#1 and Breast#2, of distinct size and interior tissue heterogeneity pattern:* evaluate the imaging system performance in cases of significantly different distances between the skin and the probes, which is a critical factor for the calibration of the mutual coupling between probes, before the scan data can be exploitable by the imaging algorithm. The two breast phantoms are depicted in **Figures 3** and **4**.

II. *Employ azimuthal rotation of each phantom placement in the scanner and repeat the measurement:* identify potential asymmetries of the MWBI scanner. Two azimuthal rotational positions of 0° and 180°, of both phantoms, have been systematically employed in the tests. In addition, all phantom configurations were rotated by 90° and 270° in a series of 16 MWBI scans, performed on a single day of the test campaign, to provide more experimental data on any potential asymmetries of the system.

III. *Evaluation using three tumour phantoms:* use of a macro-lobulated tumour phantom (Tumour#4) and two tumour phantoms of the same micro-lobulated shape model, however one tumour phantom being new and used for the first time in this test campaign (Tumour#3), while the other tumour phantom (Tumour#1) being used in numerous experimental tests before, resulting in progressive degradation of its shape sharpness and potentially its dielectric properties. It is reminded that the tumour phantoms are immersed in a chemically aggressive liquid mixture mimicking the fibro-glandular tissue of the Wavelia™ breast phantoms [32]. The immersion of the tumour in this liquid is performed each time manually using a thin string to hold it in position (additional source of variability). Photos of the three tumour phantoms, taken on the first day of the use of each, as part of the hereby reported test campaign, are shown in **Figure 3**.

IV. *Repetition of each MWBI scan twice, while the breast phantom remains idle at the exact same position*: the comparison of the achieved imaging results in each set of two tests is intended to demonstrate the expected level of reproducibility of the MWBI scan, for various experimental setups. The main sources of variability being evaluated here, necessarily as an aggregate, for the Wavelia#2 MWBI imaging system are the following:

a. the noise level in the RF measurement itself;

b. the imperfect mechanical stability of the scanner;

c. the uncertainty introduced by the imaging algorithm on its own, i.e. residual numerical instabilities and/or estimation uncertainties.

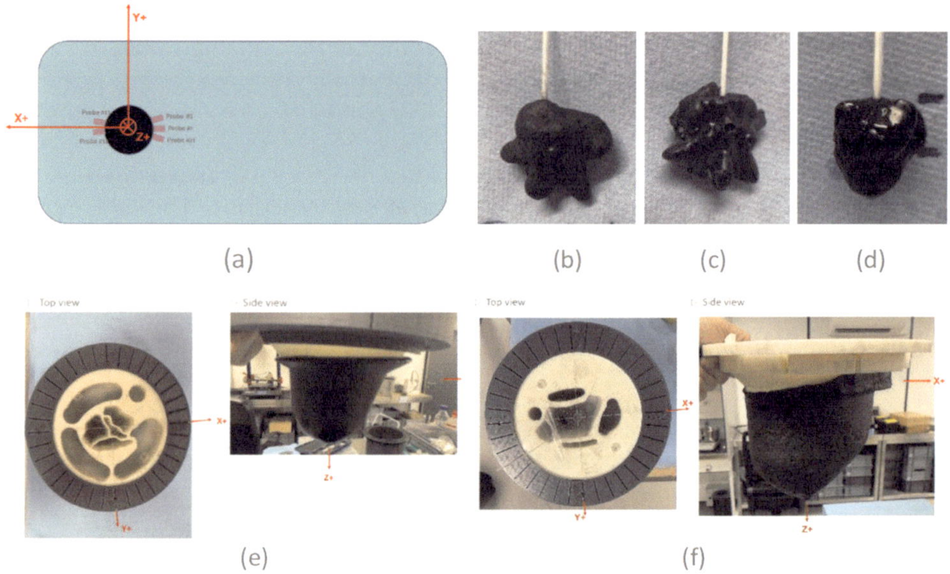

Figure 3.
(a) The Wavelia#2 examination table—schematic top view and imaging coordinate system, (b) the Tumour#1, (c) the Tumour#3, (d) the Tumour#4, (e) the Breast#2, and (f) the Breast#1 phantom.

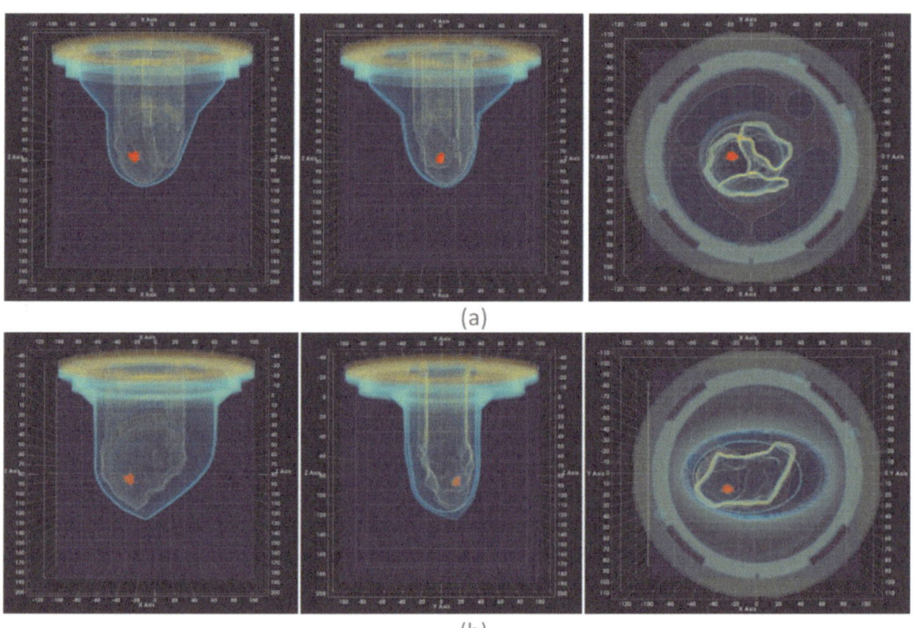

Figure 4.
The breast phantoms, with tumourous inclusion, in the Wavelia#2 imaging coordinate system: (a) Breast#2, (b) Breast#1 at the 0° rotational position.

A sequence of 40 experimental setups defined in total the on-site testing and validation campaign of the Wavelia#2 prototype at the GUH clinical investigation site. This resulted in a total number of 80 MWBI scans, given that every experimental setup was measured twice, without touching the breast phantom, and with main objective the ultimate evaluation of the repeatability of the imaging result, in various experimental conditions. The 80-MWBI scan dataset was collected during a period of 5 consecutive days (30 January – 3 February, 2023) and also on a single day approximately 1 month earlier (22 December 2022).

Each breast phantom was partially scanned in a 44 mm vertical extent around the theoretical centroid of the tumourous inclusion, at each repetition. This adds-up to 23 vertical scan positions, with a moving step of 2 mm of the probe array. That is the scan data contributing to the reconstructed breast images being part of the analysis.

4. Representative subset of imaging results for Wavelia#2 scan repeatability assessment

Imaging results for a representative subset of 34 MWBI scans are presented in **Figures 5–11**. The presented images are the ones that have also been included in the quantitative assessment of the imaging system repeatability, which is presented in the following section.

According to the methodology in [37], breast images were systematically generated for a set of four combinations for the (MAL, MPR) set of parameters defining the sensor fidelity zone, i.e. MAL = [−40–35] dB and MPR = [5 10] dB. Averaging of the images that have been generated for the multiple sensor fidelity zone settings was further performed, to cancel out ambiguities. The averaged images are presented in **Figures 5–11** and considered by default in the quantitative assessment of repeatability.

Systematic generation of the breast images for a pre-defined set of narrow and wide search ranges for the pc_fib parameter was performed, in accordance with the methodology defined in [40], towards lesion detectability assessment based on persistence. Imaging results for a fixed pc_fib search range, the best-fitting to the tumour location in each given experimental setup, i.e.

- Breast#2 phantom results – pc_fib search range = [50–70]%
- Breast#1 phantom results – pc_fib search range = [20–50]%

are presented in the following. It has been verified though and confirmed that all the presented imaging results are persistent in varying pc_fib search range and would clearly declare detectability of the visible, tumour phantom-associated, ROI.

The date on which each MWBI scan was performed is indicated in the 1st column of each set of tabulated imaging results in **Figures 5–11**. The imaging results of two scan repetitions of each experimental setup appear in the 2nd column, between which the phantom and tumour have not been moved. The experimental setup is indicated below for each couple of repeated MWBI scan results. The dynamic range of the image intensity is indicated on the colour bar at the upper part of each image. The partial transparency profile of the MWBI image colour map remains fixed in all illustrations. The contour of the mould containing the fibro-glandular tissue-mimicking liquid and the tumour phantom structure (centered at (X,Y,Z) = (20,0,77) mm in the case of the Breast#2, Rotation 180° setup in **Figure 5** and visible in purple colour) are overlaid with the MWBI images as reference.

A qualitative assessment of the results presented in **Figure 5** indicates very good levels of repeatability of the images reported by the Wavelia#2 prototype when the experimental setup remains idle (i.e. comparison of the two images per row).

22/12/2022

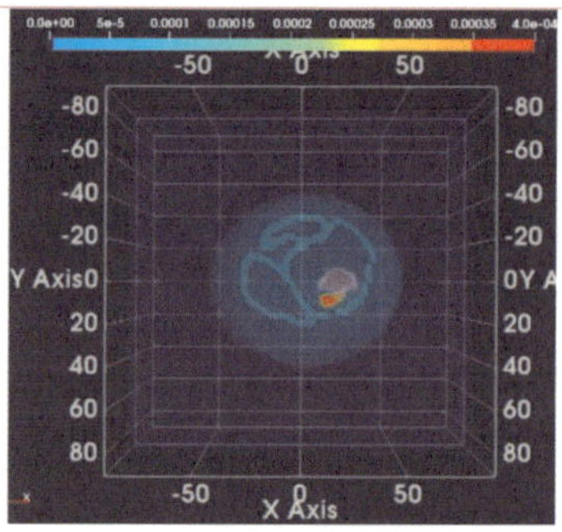

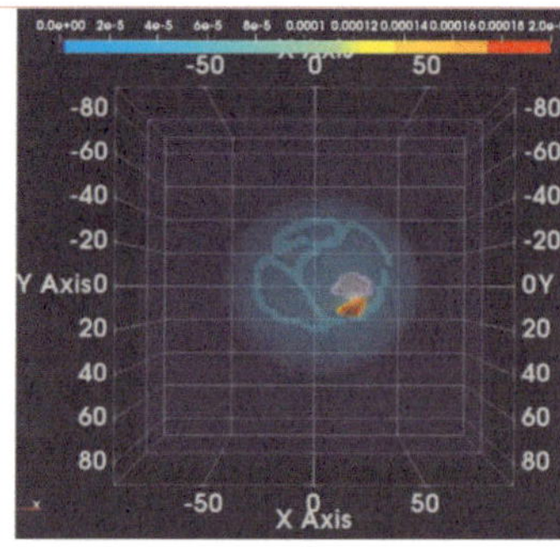

Breast#2 phantom, Rotation 180°, ***Tumour#1***

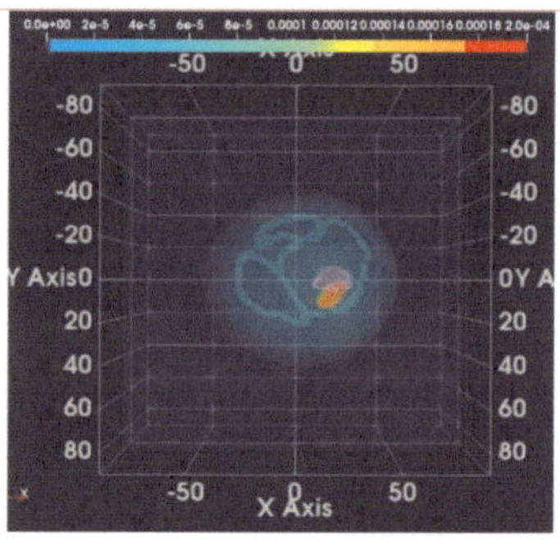

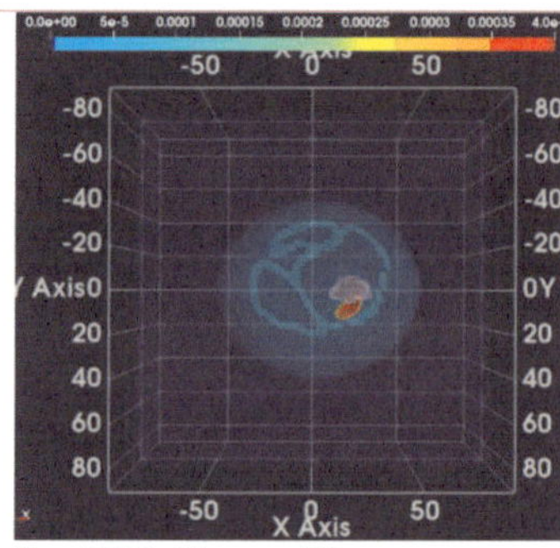

Breast#2 phantom, Rotation 180°, ***Tumour#3***

30/01/2023

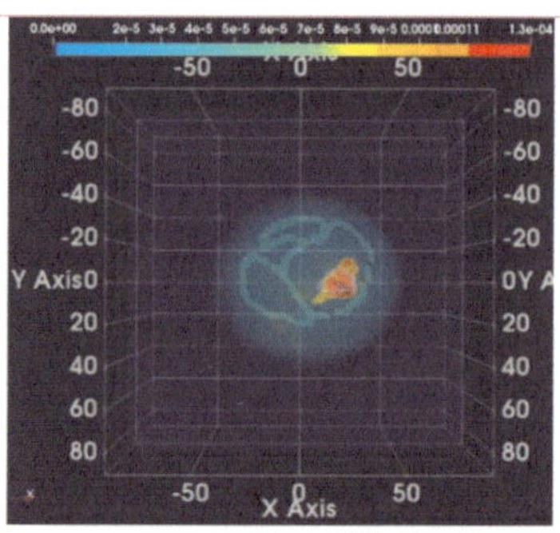

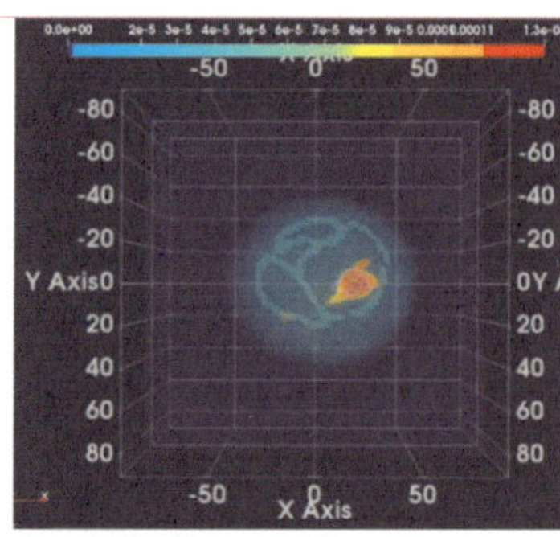

Breast#2 phantom, Rotation 180°, ***Tumour#3***

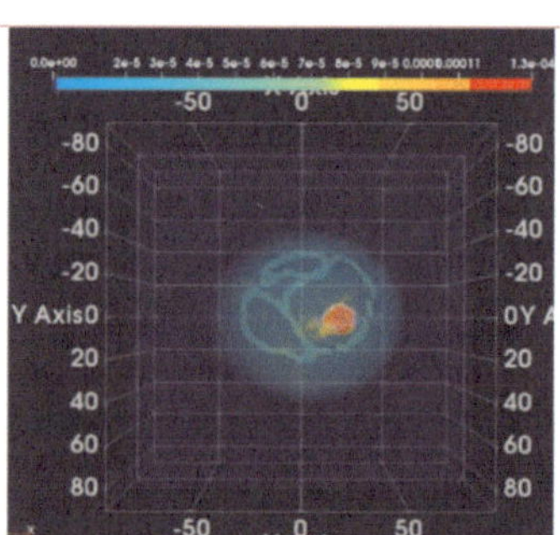

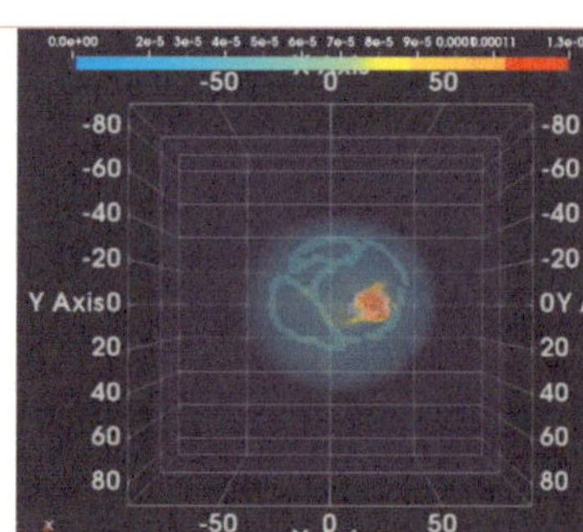

Breast#2 phantom, Rotation 180°, ***Tumour#4***

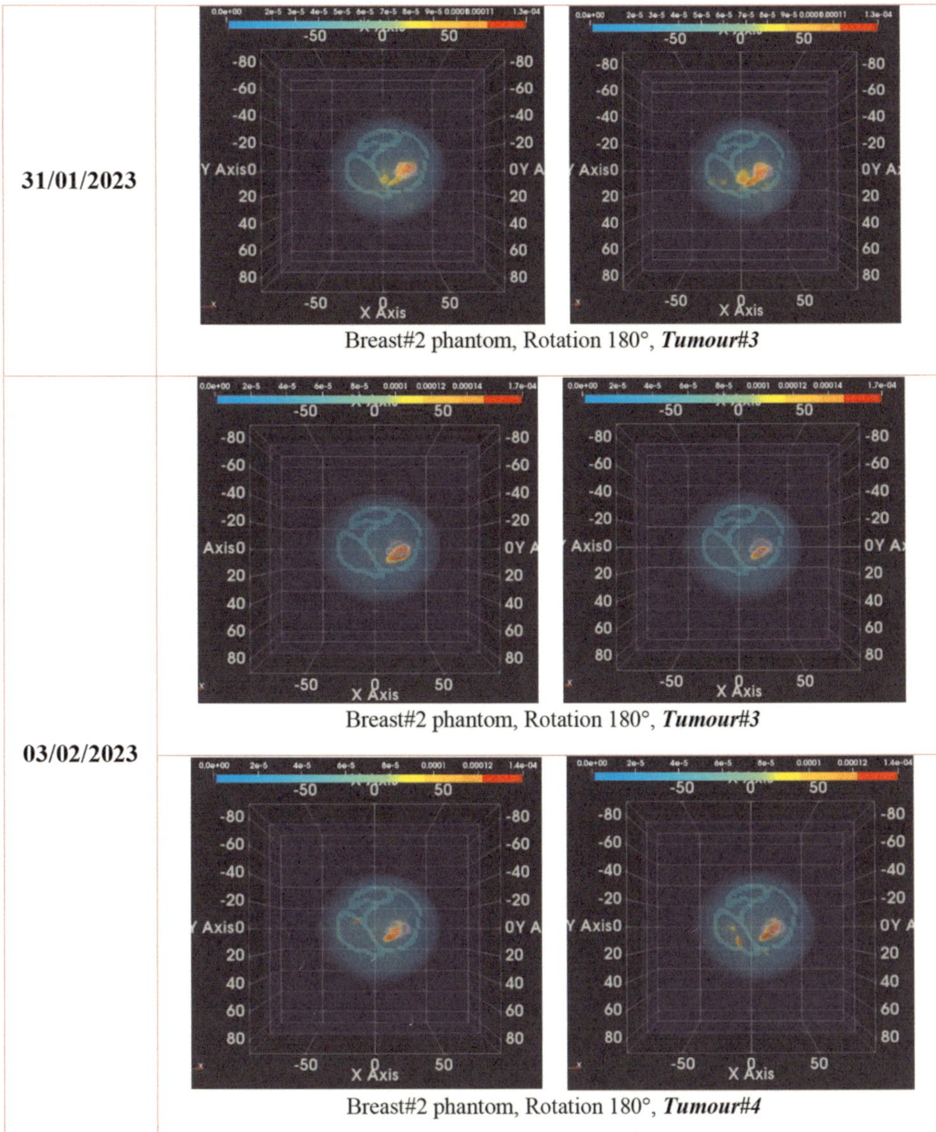

Figure 5.
Breast#2, rotation 180°: Sequence of MWBI scan results. 14 scans performed on four different days.

For the scans performed on the same day, using two different tumour models, the main source of variability is associated with the replacement and repositioning of the tumour phantom, which is performed manually using a flexible string. The following phenomena may occur during the tumour replacement, in an uncontrolled manner: (a) rotation of the tumour during/after insertion in the phantom, (b) tendency of the tumour phantom to adhere to the interior wall of the fibro-glandular tissue-mimicking mould, (c) varying immersion depth of the tumour in the breast. The breast phantom is not repositioned between two tumour exchanges on the same day.

For the scans performed on different days, aside from the aforementioned, tumour-related sources of variability, the principal source of variability is associated

DOI: http://dx.doi.org/10.5772/intechopen.1002506

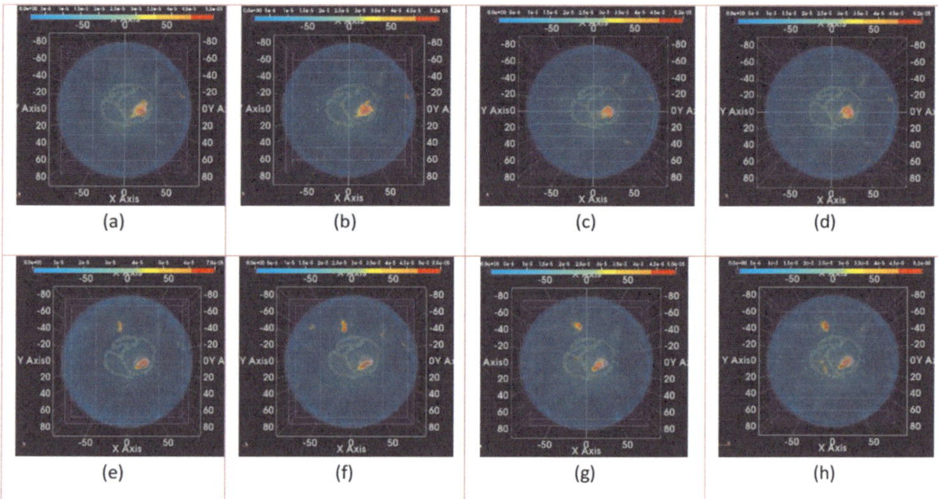

Figure 6.
Breast#2, rotation 180°: Full imaging scene reconstructions for the 4 MWBI scans of 2 days: (a)-(d) 30 January (e)-(h) 3 February.

with the potential non-optimal immersion of the phantom itself in the transition liquid of the scanner, resulting in the presence of air bubbles or other local inhomogeneities on the coverage of the skin phantom with transition liquid.

In **Figure 6**, the reconstruction of the full imaging scene within the cylindrical container (not only the image within the breast) is depicted for the four MWBI scans that were performed on the 30th January and the four scans of the 3rd of February with the Breast#2 phantom at Rotation 180°. The first two scans of each day (**Figure 6a**, **b**, **e** and **f**) were performed using Tumour#3, while the latter (**Figure 6c**, **d**, **g** and **h**) using Tumour#4.

The out-of-the-breast portion of the presented images in **Figure 6** provides a clear explanation for the visibly decreased Signal-to-Clutter Ratio (SCR) [34] on the breast phantom images of the 3rd of February (particularly the ones including the Tumour#4 phantom), when compared to the images of the 30th of January. A strong artefact radar echo appears systematically in the liquid, mostly dominant in the vicinity of the interface with the breast skin on all the four images of the 3rd of February. This is expected to be related with sub-optimal phantom placement and/or liquid consistency after mechanical injection in the scanner. The level of repeatability of the reconstructed images, both within and out-of-the breast, is remarkable, in the below 8 MWBI scans demonstration.

In **Figures 7–9**, indicative imaging results for repeatability assessment are presented at three distinct rotational positions of the Breast#2 phantom: 0°, 270° and 90° respectively. In **Figures 10–11**, a subset of indicative imaging results is presented for two rotational positions (270° and 90°) of the Breast#1 phantom.

The clarity/focusing of the image, as well as the SCR and the tumour localization accuracy, may vary from one setup to the other. This is not a surprise and can be acceptable, given the numerous sources of experimental variability. In addition, the breast phantoms are very heterogeneous, and the weak echo of the tumour phantom (low dielectric contrast against the surrounding fibro-glandular tissue) is highly sensitive to slight changes in the geometry. The tested experimental setups are of unrealistically high complexity and much discretized, compared to the real human breast.

22/12/2022	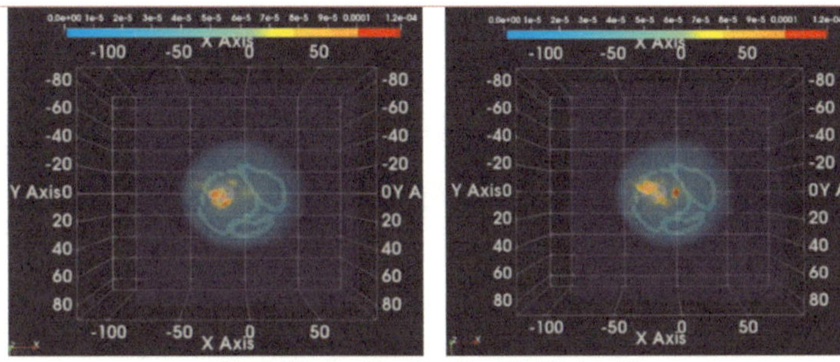 Breast#2 phantom, Rotation 0°, ***Tumour#3***
31/01/2023	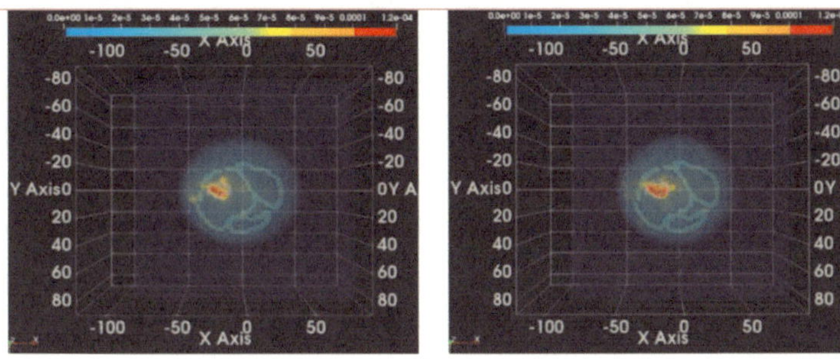 Breast#2 phantom, Rotation 0°, ***Tumour#3***

Figure 7.
Breast#2, rotation 0°: Sequence of MWBI scan results for repeatability assessment.

01/02/2023	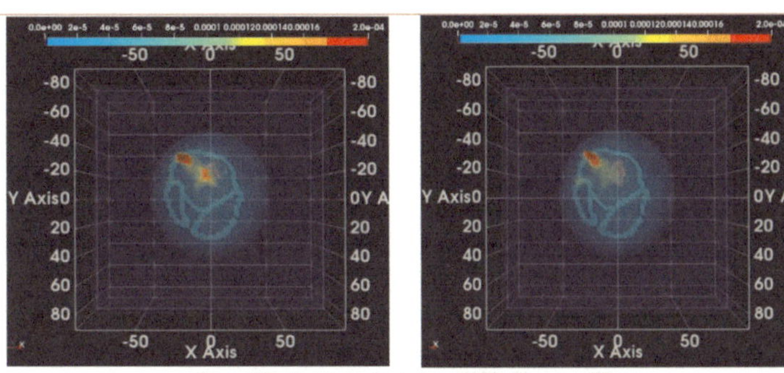 Breast#2 phantom, Rotation 270°, ***Tumour#3*** 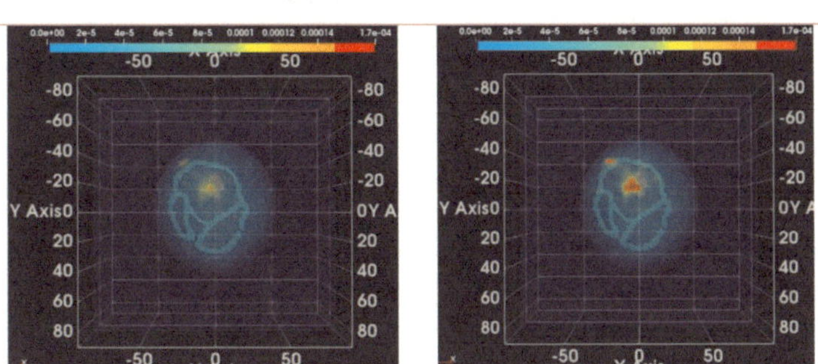 Breast#2 phantom, Rotation 270°, ***Tumour#4***

Figure 8.
Breast#2, rotation 270°: Sequence of MWBI scan results for repeatability assessment.

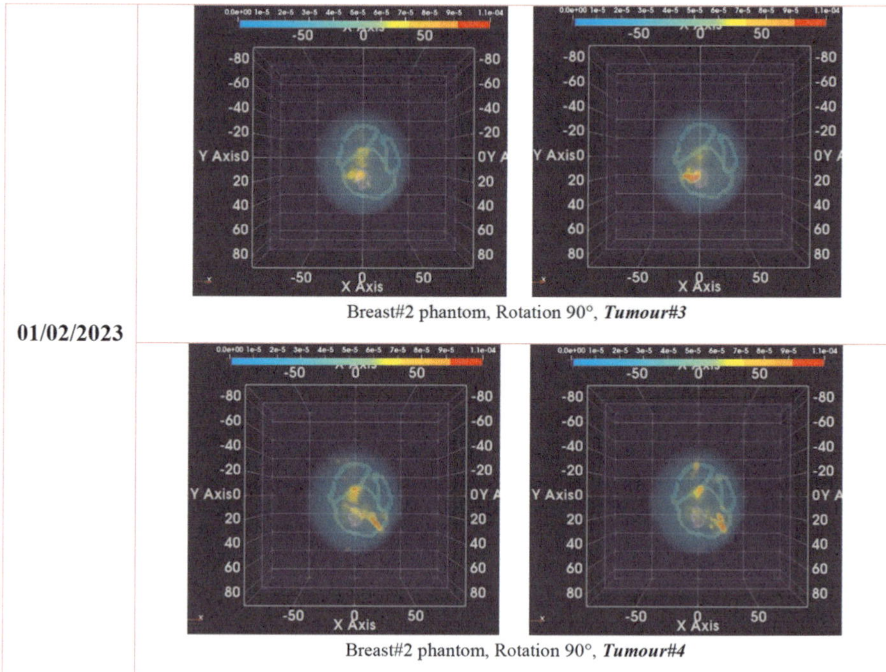

Figure 9.
Breast#2, rotation 90°: Sequence of MWBI scan results for repeatability assessment.

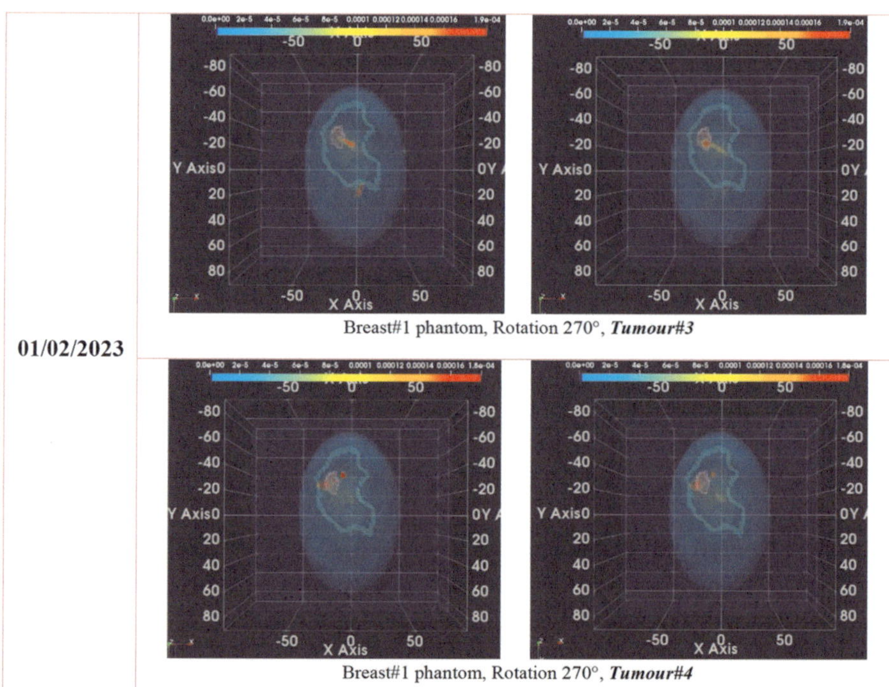

Figure 10.
Breast#1, rotation 270°: Sequence of MWBI scan results for repeatability assessment.

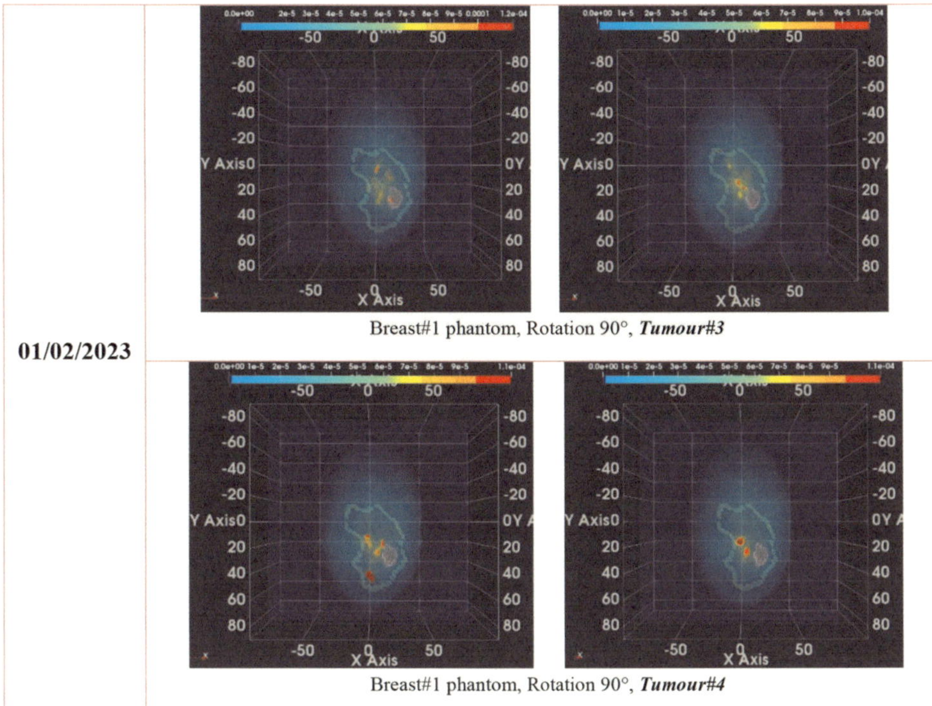

Figure 11.
Breast#1, rotation 90°: Sequence of MWBI scan results for repeatability assessment.

5. Quantitative assessment of repeatability: methodology and results

Overall, it is important for the imaging system to demonstrate repeatability in its results, so that it can be trusted with confidence to deliver similar results under slight adjustments to patient position, temperature, operational process etc. during actual patient scans. Thus, in this section, the repeatability is being quantitatively assessed for the set of 34 MWBI scan imaging results that were qualitatively assessed in the previous section. Quantitative assessment is achieved by measuring the similarity between the final 3D-images obtained using the MWBI imaging software.

To measure the similarity, the sum of squared differences has been used as a metric [42], which is defined in the following way:

$$S_{sq} = \sum_{n,m,l} \left(J_{n,m,l} - I_{n,m,l}\right)^2 \tag{1}$$

where N x M x L are the dimensions of the 3D images J and I. This has been divided by the term:

$$\sqrt{\sum\nolimits_{n,m,l} J^2_{n,m,l} \times \sum\nolimits_{n,m,l} I^2_{n,m,l}} \tag{2}$$

to provide a more appropriate scale to the [0, 1] range. More sophisticated similarity metrics can be derived using e.g. feature extractors. However, at this stage of development, it was desired to use a simple metric which directly measures

DOI: http://dx.doi.org/10.5772/intechopen.1002506

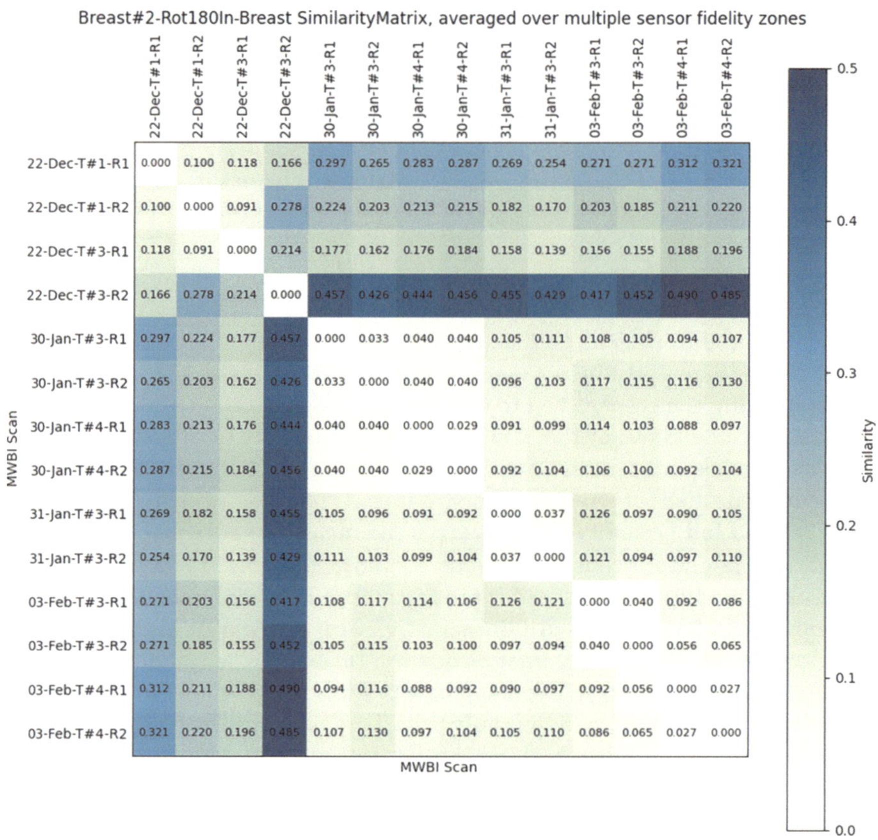

Figure 12.
In-breast similarity matrix for repeated measurements of breast #2 with rotation = 180°.

pixel-to-pixel difference. This way, we would have a more primitive, raw, estimation of the similarity between repeated measurements, without having the benefit of rotation- or scale-invariant features being used to measure similarity.

In **Figure 12**, the similarity matrix of Breast #2 with Rotation = 180° is shown, for the 14 images depicted in **Figure 5** (averaging over multiple sensor fidelity zones applied). Each element of the matrix corresponds to the similarity obtained using the sum-squared difference metric between the final 3D-images of the two respective MWBI scans. This similarity was computed from the volume inside the breast, i.e., not including the image reconstruction in the transition liquid of the cylindrical container, out-of-the breast. The following three similarity zones can be interestingly spotted in **Figure 12**.

- *L0:* this is the similarity level observed between consecutive MWBI scans performed on the same day, without moving the breast phantom or the tumour in-between the two scans (e.g. 3-Feb-T#3-R1 to 3-Feb-T#3-R2, where T# and R# stand for Tumour # and Repetition #, respectively). *L0* is meant to quantify the expected level of variability that is introduced by the imaging system itself (measurement variability + uncertainty of the imaging algorithm). L0 is systematically associated with the smallest values of difference i.e., the best similarity in **Figure 12**, as expected.

- *L1:* This zone corresponds to scans performed on the same day but involving replacement and repositioning of the tumour model in the breast phantom, for instance 30-Jan-T#3-R1 to 30-Jan-T#4-R1. The values are generally higher, thus the similarity level degraded, for the L1 zone, when compared to the L0-zone.

- *L2:* Measurements of the same experimental setup from different dates fall into this category. These values are typically higher than those of L1 but can be close.

Furthermore, it is apparent from **Figure 12**, that the scans of the 22nd of December, and T#3-R2 in particular, which appears as an outlier, are substantially more different, compared to the rest of the scans. To the contrary, the 30th of January appears to be the date with the most stable measurements, with the L1-similarity values approaching the respective L0-values.

In **Figure 13**, side views of the 3D images (XZ plane) are depicted for four MWBI scans of the same experimental setup (Breast#2 phantom, Rotation 180°, Tumour#3) performed on four distinct days. It is apparent that the reason for the decreased similarity of the 22nd December scans versus the measurements that were performed post the 30th of January is that the tumour response is detected at a slightly higher position and with a visibly increased intensity on the 22nd of December. The latter phenomenon could have to do with the sensitivity on the MWBI imaging software part, but does not affect the final image's quality, related to the tumour's detection and localization, in a negative way. The same observation holds when comparing the images in **Figure 14** as well, which involve the Tumour#1 or Tumour#4 (instead of the Tumour#3) and otherwise the same configuration as for **Figure 13**. It is clearly visible in **Figures 5, 13** and **14** that the SCR level and the overall image quality is very good for the 22-December scans, however the implementation of the experimental

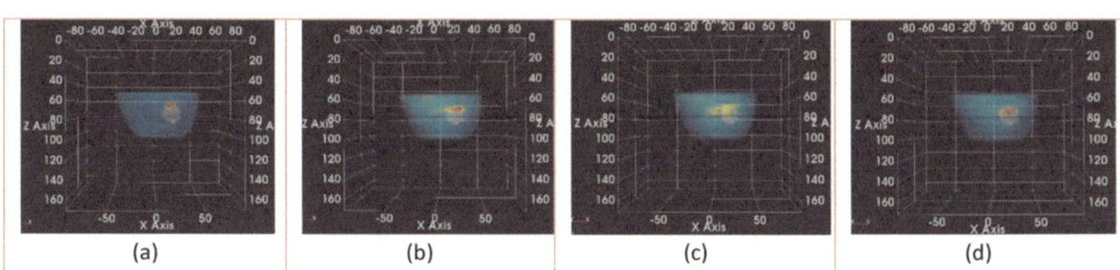

Figure 13.
Effective vertical position of the tumour phantom: Scans with Breast#2, Rotation180° and Tumour#3 inclusion: (a) 22-Dec-T#3-R2 (b) 30-Jan-T#3-R1 (c) 31-Jan-T#3-R1 (d) 03-Feb-T#3-R1.

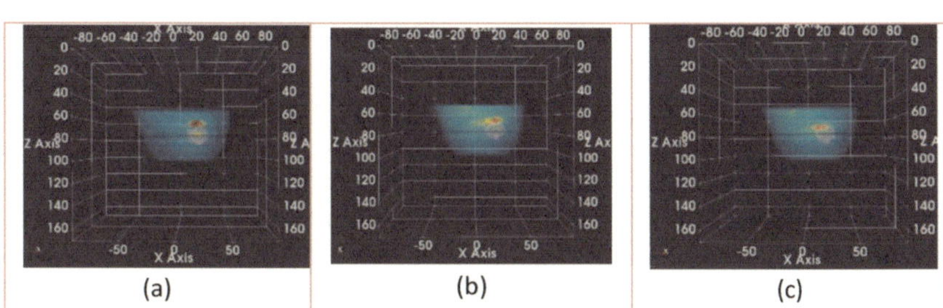

Figure 14.
Effective vertical position of the tumour phantom: Scans with Breast#2, Rotation180° and Tumour#1 or Tumour4 inclusion: (a) 22-Dec-T#1-R1 (b) 30-Jan-T#4-R1 (c) 03-Feb-T#4-R1.

setup is slightly different when compared to the posterior scans of January and February.

In addition, the increased L0 levels for the repetitions of the 22-December scans, as indicated in **Figure 12**, suggest potentially not sufficient waiting time for stabilization/immobilization of the tumour, after manual insertion in the breast phantom and before starting the 1st scan. This experimental factor was learnt from numerous preliminary tests with these breast phantoms and its handling got progressively optimized towards the end of the test campaign, when a more systematic series of MWBI scans was possible.

In **Figure 15**, the evolution of the L0-Similarity level over the full set of 34 MWBI images being considered in this analysis and shown in **Figures 5–11** is depicted. Four curves are present in the figure, for either the full volume or only the volume inside the breast and the average or a single sensor fidelity zone.

The following observations can be made about these curves:

- The full volumes follow the same tendency as the in-breast volumes and appear to have slightly degraded similarity (higher values), barring the 22-Dec-B#2-Rot180 outlier scans.

- Averaging the sensor fidelity zones does not have a significant effect on similarity.

- Looking at the in-breast similarity curves, the L0-similarity is relatively stable around the 0.04 mark, again except for the 22-Dec-B#2-Rot180 scans. This is an important outcome of this analysis, concerning the stability and repeatability of the Wavelia#2 MWBI scan and image reconstruction in the case of repetition of a fixed experimental setup.

Figure 16 depicts the respective L1-Similarity curves. It can be observed that multiple values correspond to the same date, breast and rotation, as the L1 similarity

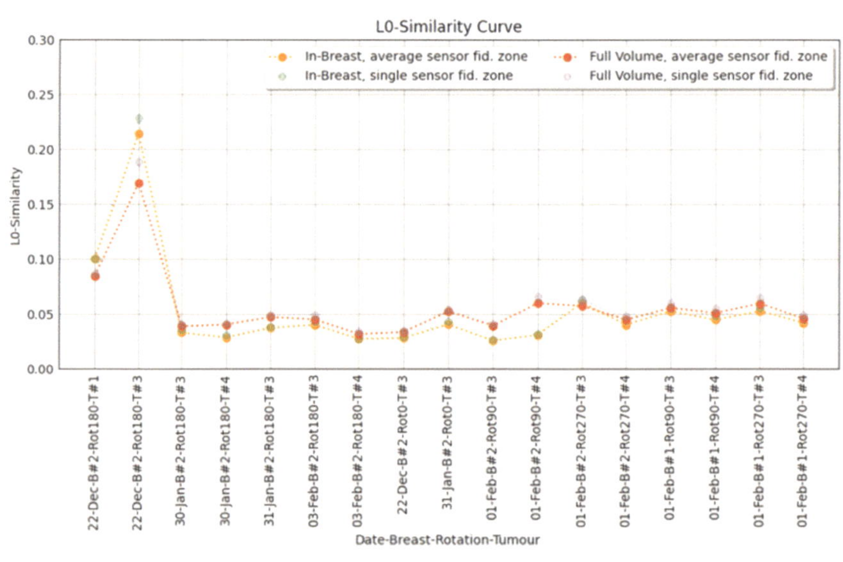

Figure 15.
L0-similarity for different dates, breast phantoms, rotations and tumours.

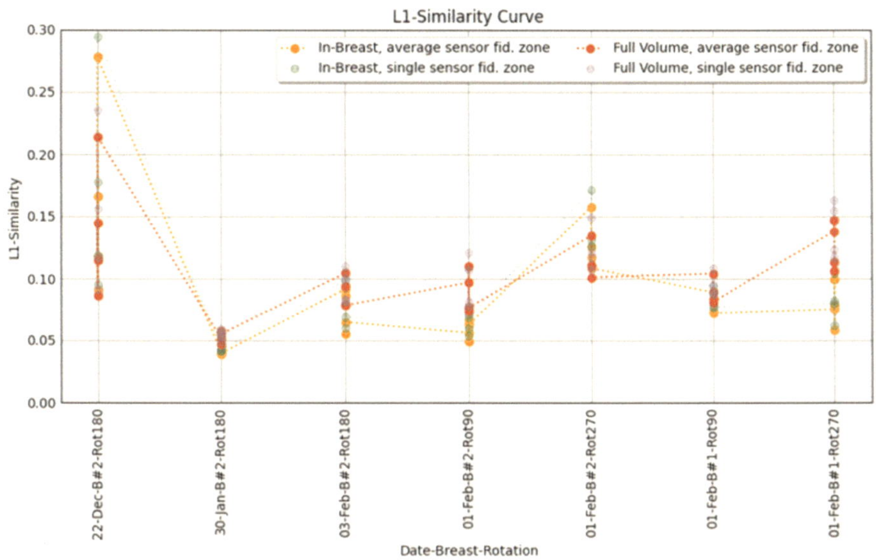

Figure 16.
L1-similarity for different date, breast phantoms and rotations.

compares measurements from the same date and the same breast phantom, after replacement and repositioning of the tumour phantom only. What can be inferred from this figure is that:

- Again, the full volumes vs. in-breast volumes, as well as single vs. average fidelity zones present the same trends as in L0.
- The values for the in-breast volume range from 0.04 (close to L0), up to 0.15. Human error is a deciding contributing factor when replacing the tumour, which can lead to similarity values 3 times as high as those of L0.

Concerning the L2-similarity levels, it can directly be observed in **Figure 12** that the L2-Similarity values can be close to L1 but are in general higher. Repeating the scan of the same experimental setup on different dates (L2) requires complete repositioning of the phantom, as well as tumour placement, which provides a margin for even greater human error. In addition, the variation and potential impact of the environmental conditions on which the MWBI scan is performed (i.e. temperature within the room and the scanner, a distinct batch of transition liquid being injected in the scanner, distinct operator preparing the setup/performing the scan) are part of the L2 evaluation. The mean and standard deviation values for all similarity zones are depicted in **Table 2**, for different configurations, as well as for the concatenated data of all of them (Total).

According to the table, the expected tendency for progressive degradation of the achieved level of repeatability when passing from L0 to L1 and L2 are indeed followed, such that L0 < L1 < L2. The similarity levels are clearly degraded (higher values of the metric defined in Eqs. (1) and (2)) when including the scans of the 22nd December, Breast#2 phantom, Rotation 180° in the analysis, as earlier discussed and experimentally justified. When these datasets are excluded from the analysis, it is

DOI: http://dx.doi.org/10.5772/intechopen.1002506

Configuration	L0		L1		L2	
	Mean	Std	Mean	Std	Mean	Std
Breast#2 Rot 180° (excl. 22-Dec)	0.033	0.005	0.057	0.02	0.104	0.01
Breast#2 Rot 180° (incl. 22-Dec)	0.069	0.064	0.092	0.067	0.2	0.119
Breast#2 Rot 0°	0.034	0.006	—	—	0.132	0.013
Total	0.061	0.058	0.092	0.067	0.196	0.117

Table 2.
In-breast volume mean and std. of different levels of similarity for different configurations.

interesting to confirm that comparable levels of the L0 similarity (as depicted and analysed in more detail in **Figure 15**), but also of the L2 similarity, are achievable on both Breast#2 Rotation 180° and Breast#2 Rotation 0° experimental setups.

6. Conclusions

Imaging results for 34 MWBI scans, being part of the on-site evaluation for acceptance of the Wavelia#2 MWBI system at the GUH clinical investigation site for a Phase-II Pilot#1 clinical investigation, have been presented. The experimental testing and validation of Wavelia#2 was performed on highly heterogeneous and complex breast phantoms, and in diverse experimental setups. Significant variability of each experimental setup itself was inherent in the MWBI scans being part of this analysis. All the experimental setups were systematically scanned twice, all along the test campaign, without moving the breast phantom or tumour in-between the two consecutive MWBI scans. The main objective of this test campaign was to understand and assess the sources of variability, in an attempt to isolate and quantify the impact of the imaging system itself (measurement variability and imaging algorithm uncertainty) on the repeatability of the MWBI imaging results.

The similarity between the images of repeated scans under different conditions of repetition was quantitatively assessed using the sum-squared difference metric. It was observed both qualitatively/visually and quantitatively that repeating a measurement without changing anything in the experimental setup (L0 level) yields very similar imaging results. A quite stable level of image similarity was demonstrated to be achievable for the L0 level, on multiple tests with highly diverse experimental setups. The inherent variability introduced by each distinct implementation of the experiment was demonstrated to be mapped to L1 and L2 similarity levels higher than the L0 levels, however consistently reproduced on multiple days of testing and with diverse experimental setups being employed. The reported results are promising, in terms of the expected performance of the Wavelia#2 prototype in the ongoing clinical investigation, the reliability of the reported images, as well as towards suitability of the prototype for a repeatability study on patients.

Author details

Angie Fasoula*, Petros Arvanitis and Luc Duchesne
Wavelia Healthcare, MVG Industries, Villejust, France

*Address all correspondence to: angie.fasoula@mvg-world.com

References

[1] Fear EC, Stuchly MA. Microwave detection of breast cancer. IEEE Transactions on Microwave Theory and Techniques. 2000;**48**(1):1854-1863

[2] Nikolova NK. Microwave imaging for breast cancer. IEEE Microwave Magazine. 2011;**12**(7):78-94

[3] Conceição RC, Mohr JJ, O'Halloran M. An Introduction to Microwave Imaging for Breast Cancer Detection. Biological and Medical Physics. 2016. ISBN: 978-3-319-27865-0

[4] Gabriel S, Lau RW, Gabriel C. The dielectric properties of biological tissues: III. Parametric models for the dielectric spectrum of tissues. Physics in Medicine and Biology. 1996;**41**(11):2271-2293

[5] Campbell AM, Land DV. Dielectric properties of female human breast tissue measured in vitro at 3.2 GHz. Physics in Medicine and Biology. 1992;**37**(1): 193-210

[6] Preece AW, Craddock I, Shere M, Jones L, Winton HL. MARIA M4: Clinical evaluation of a prototype ultrawideband radar scanner for breast cancer detection. Journal of Medical Imaging. 2016;**3**(2):033502

[7] Meaney PM, Kaufman PA, Muffly LS, Click M, Poplack SP, Wells WA, et al. Microwave imaging for neoadjuvant chemotherapy monitoring: Initial clinical experience. Breast Cancer Research. 2013; **15**(2):R35

[8] Zastrow E, Davis SK, Hagness SC. Safety assessment of breast cancer detection via ultrawideband microwave radar operating in pulsed-radiation mode. Microwave and Optical Technology Letters. 2007;**49**(1):221-225

[9] Keemers-Gels ME, Groenendijk RP, van den Heuvel JH, Boetes C, Peer PG, Wobbes TH. Pain experienced by women attending breast cancer screening. Breast Cancer Research and Treatment. 2000;**60**(3):235-240

[10] O'Loughlin D, O'Halloran MJ, Moloney BM, Glavin M, Jones E, Elahi MA. Microwave breast imaging: Clinical advances and remaining challenges. IEEE Transactions on Biomedical Engineering. 2018;**65**(11): 2580-2590

[11] Kwon S, Lee S. Recent advances in microwave imaging for breast cancer detection. International Journal of Biomedical Imaging. 2016;**2016**:5054912

[12] Moloney BM, O'Loughlin D, Elwahab SA, Kerin MJ. Breast cancer detection—A synopsis of conventional modalities and the potential role of microwave imaging. Diagnostics. 2020; **10**(2):103

[13] Benny R, Anjit TA, Mythili P. An overview of microwave imaging for breast tumour detection. Progress in Electromagnetics Research. 2020;**87** (January):61-91

[14] Aldhaeebi MA, Alzoubi K, Almoneef TS, Bamatra SM, Attia H, Ramahi OM. Review of microwaves techniques for breast cancer detection. Sensors (Switzerland). 2020;**20**(8):1-38

[15] Shere M, Lyburn I, Sidebottom R, Massey H, Gillett C, Jones L. MARIA®M5: A multicentre clinical study to evaluate the ability of the Micrima radio-wave radar breast imaging system (MARIA®) to detect lesions in the symptomatic breast. European Journal of Radiology

[Internet]. 2019;**116**:61-67. Available from:. DOI: 10.1016/j.ejrad.2019.04.017

[16] Sani L, Ghavami N, Vispa A, Paoli M, Raspa G, Ghavami M, et al. Novel microwave apparatus for breast lesions detection: Preliminary clinical results. Biomedical Signal Processing and Control. 2019;**52**:257-263

[17] Sani L, Vispa A, Loretoni R, Duranti M, Ghavami N, Sánchez-Bayuela DA, et al. Breast lesion detection through MammoWave device: Empirical detection capability assessment of microwave images' parameters. PLoS One. 2021;**16**(4):e0250005

[18] Rana SP, Dey M, Loretoni R, Duranti M, Ghavami M, Dudley S, et al. Radiation-free microwave technology for breast lesion detection using supervised machine learning model. Tomography. 2023;**9**(1):105-129

[19] Janjic A, Cayoren M, Akduman I, Yilmaz T, Onemli E, Bugdayci O, et al. SAFE: A novel microwave imaging system design for breast cancer screening and early detection—Clinical evaluation. Diagnostics. 2021;**11**(3):533

[20] Janjic A, Akduman I, Cayoren M, Bugdayci O, Aribal ME. Gradient-boosting algorithm for microwave breast lesion classification—SAFE clinical investigation. Diagnostics. 2022;**12**(12):3151

[21] Janjic A, Akduman I, Cayoren M, Bugdayci O, Aribal ME. Microwave breast lesion classification – Results from clinical investigation of the SAFE microwave breast cancer system. Academic Radiology. 20 Dec 2022: S1076-6332(22)00641-9

[22] Meaney PM, Fanning MW, Zhou T, Golnabi A, Geimer SD, Paulsen KD. Clinical microwave breast imaging - 2D results and the evolution to 3D. In: Proceedings of the 2009 International Conference on Electromagnetics in Advanced Applications, ICEAA'09. Turin, Italy: IEEE; 2009. pp. 881-884

[23] Moloney BM, McAnena PF, Abd Elwahab SM, Fasoula A, Duchesne L, Gil Cano JD, et al. Microwave imaging in breast cancer – Results from the first-in-human clinical investigation of the Wavelia system. Academic Radiology. Jan 2022;**29**(Supplement 1):S211-S222

[24] Moloney BM, McAnena PF, Elwahab SM, Fasoula A, Duchesne L, Gil Cano JD, et al. The Wavelia microwave breast imaging system – Tumour discriminating features and their clinical usefulness. The British Journal of Radiology. 2021;**94**(1128):20210907 [Epub ahead of print]

[25] Oliveira BL, O'Loughlin D, O'Halloran M, Porter E, Glavin M, Jones E. Microwave breast imaging: Experimental tumour phantoms for the evaluation of new breast cancer diagnosis systems. Biomedical Physics & Engineering Express. 2018;**4**(2):025036

[26] O'Loughlin D, Oliveira BL, Santorelli A, Porter E, Glavin M, Jones E, et al. Sensitivity and specificity estimation using patient-specific microwave imaging in diverse experimental breast phantoms. IEEE Transactions on Medical Imaging. 2019; **38**(1):303-311

[27] Joachimowicz N, Duchêne B, Conessa C, Meyer O. Anthropomorphic breast and head phantoms for microwave imaging. Diagnostics. 2018;**8**(4):85

[28] Khalid B, Khalesi B, Ghavami N, Sani L, Vispa A, Badia M, et al. 3D Huygens principle based microwave imaging through MammoWave device: Validation through phantoms. IEEE Access. 2022;**10**:106770-106780

[29] Burfeindt MJ, Colgan TJ, Mays RO, Shea JD, Behdad N, Van Veen BD, et al. MRI-derived 3-D-printed breast phantom for microwave breast imaging validation. IEEE Antennas and Wireless Propagation Letters. 2012;**11**:1610-1613

[30] Joachimowicz N, Conessa C, Henriksson T, Duchêne B. Breast phantoms for microwave imaging. IEEE Antennas and Wireless Propagation Letters. 2014;**13**:1333-1336

[31] Conceicao RC, O'Halloran M, Capote RM, Ferreira CS, Matela N, Ferreira HA, et al. Development of breast and tumour models for simulation of novel multimodal PEM-UWB technique for detection and classification of breast tumours. In: 2012 IEEE Nuclear Science Symposium and Medical Imaging Conference Record (NSS/MIC), Anaheim, CA, USA. 2012. pp. 2769-2772

[32] Fasoula A, Jg B, Robin G, Duchesne L. Elaborated breast phantoms and experimental benchmarking of a microwave breast imaging system before first clinical study. In: 12th European Conference on Antennas and Propagation (EuCAP 2018). London, UK: IEEE; 2018. pp. 1-5

[33] Fasoula A, Duchesne L, Gil Cano J, Lawrence P, Robin G, Bernard JG. On-site validation of a microwave breast imaging system, before first patient study. Diagnostics. [Internet]. 2018;**8**(3): 53. Available from: http://www.mdpi.com/2075-4418/8/3/53

[34] Porter E, O'loughlin D. Pathway to demonstrating clinical efficacy of microwave breast imaging: Qualitative and quantitative performance assessment. IEEE Journal of Electromagnetics, RF and Microwaves in Medicine and Biology. 2022;**6**(4): 439-448

[35] Fasoula A, Moloney BM, Duchesne L, Cano JDG, Oliveira BL, Bernard JG, et al. Super-resolution radar imaging for breast cancer detection with microwaves: The integrated information selection criteria. In: 41st Annual International Conference of the IEEE Engineering in Medicine & Biology Society (EMBC). Berlin, Germany: IEEE; 2019. pp. 1868-1874

[36] Devaney AJ. Time reversal imaging of obscured targets from multistatic data. IEEE Transactions on Antennas and Propagation. 2005;**53**(5): 1600-1610

[37] Fasoula A, Duchesne L, Abdoush Y, Baracco JM. Frequency-dependent, configurable, sensor fidelity zone for microwave breast imaging: System dimensioning and image quality enhancement. In: IEEE Conference on Antenna Measurements & Applications (CAMA). Antibes Juan-les-Pins, France: IEEE; 2021. pp. 487-492

[38] Lazebnik M, Popovic D, McCartney L, Watkins CB, Lindstrom MJ, Harter J, et al. A large-scale study of the ultrawideband microwave dielectric properties of normal, benign and malignant breast tissues obtained from cancer surgeries. Physics in Medicine and Biology. 2007; **52**(20):6093-6115

[39] Sugitani T, Kubota SI, Kuroki SI, Sogo K, Arihiro K, Okada M, et al. Complex permittivities of breast tumour tissues obtained from cancer surgeries. Applied Physics Letters. 2014;**104**(25): 253702

[40] Fasoula A, Duchesne L, Cano JDG, Moloney BM, Elwahab SMA, Kerin MJ. Automated breast lesion detection and characterization with the wavelia microwave breast imaging system: Methodological proof-of-concept on

first-in-human patient data. Applied Sciences. 2021;**11**(21):9998

[41] Tanaka T, Mandic DP. Complex empirical mode decomposition. IEEE Signal Processing Letters. 2007;**14**(2): 101-104

[42] Hisham MB, Yaakob SN, Raof RAA, Nazren ABA. Embedded NMW. Template matching using sum of squared difference and normalized cross correlation. In: 2015 IEEE Student Conference on Research and Development, SCOReD 2015. Kuala Lumpur, Malaysia: IEEE; 2015. pp. 100-104

Chapter 4

Dielectric Spectroscopy for the Non-Destructive Characterization of Biomaterials: Fundamentals, Techniques, and Experimentations

Nuria Novas Castellano, Dalia El Khaled, Rosa Maria Garcia, Francisco Portillo, Manuel Fernadez-Ros and Jose Antonio Gazquez

Abstract

This chapter provides an overview of research on the dielectric properties of fresh food and their applications in assessing food quality and freshness. Non-destructive methods, including dielectric techniques such as dielectric spectroscopy (DS) and bioelectrical impedance spectroscopy (BIS), have gained importance in assessing food quality without damaging the products. The importance of external appearances, such as color, size, brightness, and hardness, in determining the freshness of fruits and vegetables is emphasized. Several dielectric techniques, such as impedance, capacitance, and electrical conductivity measurements, are studied to assess quality at distinct stages of the supply chain. These techniques can detect defects, diseases, and mechanical damage and facilitate storage quality control and processing quality evaluation. Accurate measurements and instrumentation advancements are crucial for effectively implementing these techniques. The study of dielectric properties offers promising prospects for evaluating food quality and ensuring freshness. Further research and technological advances in this field can enhance the monitoring and maintaining optimal conditions for fresh produce throughout the food supply chain, reducing food waste and improving consumer satisfaction.

Keywords: dielectric properties, dielectrics, non-destructive, bioelectrical impedance spectroscopy (BIS), characterization of biomaterials

1. Introduction

The rapid advancement of technology has broadened its field of application, improving the daily lives of users and their food. Currently, research is being done on systems that will enhance food production to meet the high demand, in addition to studying its quality and improvements in transport conditions [1]. The current trend is to achieve a healthy diet, to which seasonal fruit and vegetable products

IntechOpen

significantly contribute [2]. Fresh foods are essential in the human diet due to their high nutrients, vitamins, minerals, and fiber content. The quality of these foods can be evaluated in several ways, considering their external appearances, such as color, size, brightness, smell, and hardness, or internal such as taste, texture, water content, and nutritional content. The external appearance is considered an essential criterion to consider some fruit as "fresh" and implies creating expectations about its internal quality. Manufacturers seek a balance between the appearance and maturity of the fruit in the highest qualities, such as color and hardness for consumption, in addition to achieving maximum economic performance in its marketing, including freshness and durability. In the commercialization phase, durability is essential for its processing and commercialization until it arrives at homes in optimal conditions of internal and external quality.

Before marketing, the transfer of fruits and vegetables must be controlled from the orchard to the distribution warehouses and the shops for sale. Depending on the product and the distance, the product must be moved in refrigerated chambers, where temperature and humidity affect fruits and vegetables differently. If it is not moved at the proper temperature, it causes cold damage, not arriving in the right conditions optimal for shops. Therefore, cold damage is investigated, looking for combinations of products with a wide temperature range to be transported with others of a more demanding range but of more excellent economic value [1].

Horticultural engineering has evolved a lot, and higher quality products are increasingly sought, but at the same time more resistant to transport, production, and storage conditions. More research is being conducted on post-harvest treatment, where innovative technologies make it possible to extend the useful life of the products so that they arrive in the best conditions at the consumers' tables. Constant product quality monitoring is essential at every stage of the supply chain, and it is crucial to utilize portable tools in industrial and field environments [3].

The tendency to consume high-quality fresh products has motivated the producers and marketers of these products to encourage the study of quality detection methods, optical, acoustic, and mechanical systems have been developed to assess quality based on sensory perceptions such as sight, smell, and touch [4]. Quality parameters such as nutritional value, health, and safety are less tangible for consumers because they require measurements. The results of these measurements are shown to consumers on their labels and are a claim for the most demanding consumers or who seek high-quality products in a healthy diet.

In the last decade, non-destructive methods are advancing evaluating food quality since they can be assessed without destroying the examined product. Within non-destructive methods, the main application is those based on images [5], thermal imaging [6], and spectroscopy [7]. However, many systems are based on the simulation of smell, such as the electronic nose or taste, the electronic tongue [8], and sound as acoustic systems [9].

Among other methods of assessing quality, one can consider the measurement of dielectric properties, which refer to the ability of materials to store and transmit electrical energy in an electrical field [10]. The technique known as dielectric spectroscopy (DS) refers to the interaction between a material and an externally applied electric field. In the case of fruits and vegetables, non-destructive methods such as bioelectrical impedance spectroscopy (BIS) are utilized for quality control purposes [11]. Measurement of dielectric properties offers helpful insights into the quality and freshness of food at different stages of the food chain, as well as the presence of defects or diseases [12], storage quality control, and processing quality evaluation

DOI: http://dx.doi.org/10.5772/intechopen.1002493

[13]. The measurement of these properties can be done using different dielectric techniques, such as DS (which measures the dielectric response of materials at varying frequencies), electrical impedance (a method used to measure the electrical resistance of materials), capacitance (a technique that measures the electrical energy storage capacity of materials) and electrical conductivity (a technique that assesses the ability of materials to conduct electrical current).

DS determines the dielectric properties utilizing a multifrequency impedance analyzer, which observes the electrical response when current passes through the tissue for certain measurement conditions. The approach entails the assessment of the dielectric constant and dielectric loss, which indicate the material's ability to store and discharge electrical energy. Electric permittivity encompasses the interaction between electromagnetic waves and substances, as well as the determination of charge density when subjected to an external electric field [14]. The first documentation of dielectric properties for grains was reported by Nelson in 1965. Permittivity depends on the dielectric constant and a loss factor [15], explained in the next section for varied materials ranging from solids to liquids and gases.

At present, the dielectric characteristics of many foods are known, being one of the factors that affect the measure of moisture content. Also, certain mineral substances and organic acids have the potential to undergo dissociation, producing elevated electrolytic conductivity in food.

This review presents an overview of the current research on fresh food's dielectric properties. It delves into the various dielectric techniques employed to measure these properties and explores their applications in evaluating food quality and freshness.

2. Techniques

2.1 Theory

For decades, many industrial applications have succeeded in taking advantage of the potential discovery of radio frequency to establish empirical correlations between the physical properties in materials and the dielectric properties and develop techniques for the rapid and non-destructive study of physical properties. Measurement, modeling, and applications of the dielectric properties is a unique scientific journey in dielectrics that reveals as many challenges as rewards [16].

By dielectric properties of materials are designated the electric characteristics determining the interaction of materials with electric fields. In this concern, a particular approach defines the interaction of the electric field component of the electromagnetic waves when it comes to food materials, agricultural products, and other dielectric materials.

Being dependent on the composition of material, frequency, and temperature, dielectric properties have been opted as bases for developing sensors. These rapid and nondestructive methods can be valid for assessing the physical properties of materials [17].

In discussing the applications of dielectric properties, some simplified definitions are helpful. Primarily, the propagation of electromagnetic energy in free space at the velocity of light (c) is the fundamental characteristic of all forms of electromagnetic energy. The electromagnetic parts of a material have a markable influence on the speed of propagation (V) of electromagnetic energy in that material, and this velocity (V) is given by Eq. (1), such as the magnetic permeability is given by μ and the electric permittivity by ε.

$$V = \frac{1}{\sqrt{\mu\varepsilon}} \tag{1}$$

The same equation defined in free space becomes as follows in Eq. (2), where μ_0 is the vacuum permeability y of free space, and ε_0 is vacuum permittivity.

$$V = \frac{1}{\sqrt{\mu_0\varepsilon_0}} \tag{2}$$

To interpret the molecular mechanism, the interaction of electromagnetic waves with any material is studied by Maxwell's equation [18]. The relative complex permittivity (ε^*) is represented in Eq. (3), where ε' and ε'' are commonly called the dielectric constant and loss factor, respectively, and $j = \sqrt{-1}$.

$$\varepsilon^* = \varepsilon' - j\varepsilon'' \tag{3}$$

Relatively to free space, the relative complex permittivity (ε_r^*) will be presented as follows in Eq. (4).

$$\varepsilon_r^* = \frac{\varepsilon_a}{\varepsilon_0} = \varepsilon'_r - j\varepsilon''_r \tag{4}$$

Where ε_0 is 8854×10^{-12} F/m, and ε'_r, and ε''_r are the relative dielectric constant and loss factor respectively; hence, this latter the two quantities of interest that are the real part of the complex permittivity ε referred to as the dielectric constant and the complex part given by (ε referred to as loss factor).

As a material classified as "dielectric", this reflects its ability to store energy under the application of an external electric field. When subjected to an electric field, the real part, ε', describes the ability of a material to store energy and influences the electric field distribution and the phase of waves traveling through the material; its value is related to chemical structure and intermolecular interaction [19].

The amount of energy lost or dissipated by the material under an external electric field or in any polarization mechanism generating heat is measured by the loss factor, which is the imaginary part of the permittivity [20]. Nevertheless, this loss factor is always proportional to the amount of conversion of thermal energy in food [21].

The loss factor is always positive and has smaller values than the dielectric constants. The values of dielectric constants are usually higher than the loss factor that, is commonly limited and positive and is returned to the energy dissipation such that a material is lossless when this loss factor is null [22].

Relaxation time (τ) measures the mobility of the molecules (dipoles) in a material. It is the time required for a displaced system aligned in an electric field to return to 1/e of its random equilibrium value. It is also called the time needed for dipoles to become oriented in an electric field.

The alternating electric field is slow enough for the dipoles to keep pace with the field variations at frequencies below the relaxation frequency. Since the polarization can fully develop, the loss factor, $\varepsilon r''$, is directly proportional to the frequency. As the frequency increases, $\varepsilon r''$ continues to increase, but $\varepsilon r'$ decreases due to the phase lag between the dipole alignment and the electric field. Above the relaxation frequency, both $\varepsilon r''$ and $\varepsilon r'$ decrease as the electric field is too fast to affect the dipole rotation, causing the orientation polarization to disappear.

DOI: http://dx.doi.org/10.5772/intechopen.1002493

Figure 1 shows the dielectric properties of distilled water at 25°C [23]. Many liquids and solids have been experimented with through the various models of Debye. Thermal agitation is the most significant in liquids where thermal agitation is maximal, thus producing more considerable losses at higher frequencies [24].

It is crucial to obtain the optimum frequency range at which the material under test will have appropriate dielectric properties to develop a proper heating process based on electromagnetic energy [25]. The interaction between the material and high-frequency electromagnetic energy is crucial for defining the desired properties. Consequently, it is essential to carefully select food and meals to facilitate MW propagation [26–28].

2.2 Measurement systems

Building measurement systems rely on providing equipment for the abovementioned properties. Through the past years, worldwide experiments have witnessed a variety of technologies implemented for that purpose. The built systems differ according to the material under test and the properties such as resonant frequency and quality factor from another perspective. Knowing the physical dimensions of the material and measuring the transmission and/or reflection can characterize the permittivity and permeability. At low frequencies, impedance analyzers and LCR were considerably used by simulating the material with an AC source and monitoring the capacitance and dissipation factors [27]. The impedance data analysis starts with two complex yet critical steps. The first stage involves investigating the physical circuit and estimating the necessary equations. The mathematical methods are evaluated at a second stage, allowing the model parameters to be extracted [29].

Depending on the material's nature, the fixture's choice was made so that the sample holder would fit the solid, liquid, powder, or gaseous nature of the tested material, along with appropriate software modeling the interaction of the fixture with the said material. The impedance measurement method is commonly applied by

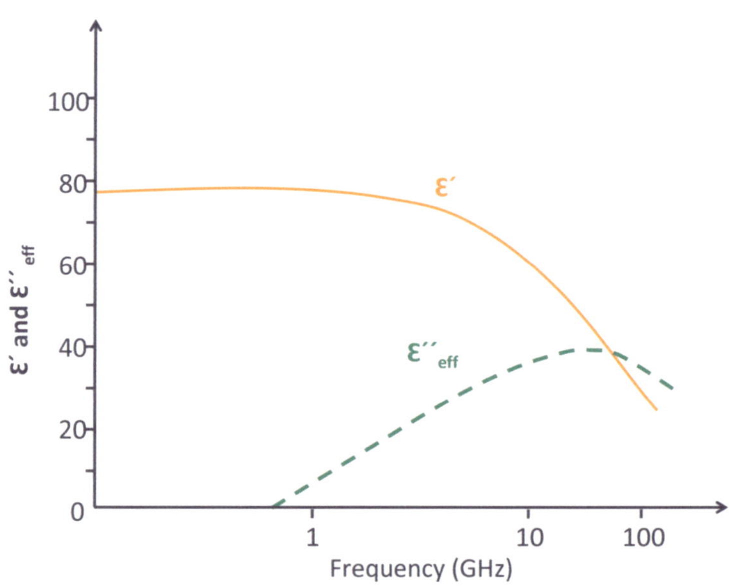

Figure 1.
The frequency-dependent dielectric characteristics of water.

applying a current across electrodes and observing the voltage. Another widespread practice for many applications is the self-balancing bridges developed by Agilent. Such bridges under a higher voltage can give a better signal-to-noise ratio higher than 10–300 mV. Such applications are robust, simple, and more practical than the old manual bridges. With advances in experiments comes the urge for more complex systems to give accurate results and cover more material. This led to a fast improvement in techniques. Moreover, high microwave frequencies applied during experiments were to produce radiation loss, especially considering the expensive design of a microwave network analyzer. Throughout the past decades, a variety of equipment has been accessible in the market for impedance measurements. However, network analyzers emerged as the most widely employed option due to their capacity to handle large electrode systems and cover a wide frequency range. Notably, vector network analyzers like PNS, PNA-L, ENA, and ENA-L are capable of sweeping high-frequency stimulus responses, ranging from 300 kHz to 110 GHz or even 324 GHz. Impedance and scalar network analyzers [30] are known for their cost-effectiveness compared to Vector Network Analyzers (VNAs). However, their usage is restricted to the frequency range associated with the α and β dispersions.

The experimental setup involves utilizing the open-ended probe technique, which includes a coaxial probe and dedicated software, to conduct a range of experiments using the network analyzer. Keysight Technologies came out with spectrum and network analyzers with high measurement integrity.

The S11 reflection parameter determines the surface of contact between the probe and the tested material using the VNA that is meant to be a probe with specialized software. The phase shift corresponding to the change in the magnitude ratio is called S11. The illustrations of reflection and transmission probes are highlighted in **Figure 2**. Some exemplary models are incorporated into the supplied software to collect the real and imaginary parts of the dielectric permittivity.

For less expensive solutions, the 85070E probe is replaced with another subminiature version (SMA) coaxial connector that can perform correctly at some good frequency ranges [32].

Conversely, coaxial transmission lines or waveguides are employed to evaluate the S21 parameter, which represents the ratio between the transmitted signal through the material under test and the incident signal. The magnetic permeability and dielectric permittivity analysis determine the S21 parameter [33].

In the reflective operating mode, the VNA produces a sinusoidal signal spanning a wide frequency range from 20 kHz to 8 GHz. The incident and reflected signal separation happen at this point [34]. From the reflected signal of the probe, dielectric properties obtained are chargeable to derive the phase and the amplitude. Theoretical models are also used to derive the frequency spectra for permittivity factors $\varepsilon'(f)$ and $\varepsilon''(f)$. Moreover, some elements should be counted as a distribution parameter system for modeling purposes, especially since the real part of $\varepsilon^*(f)$ requires applying high frequencies [35].

2.3 Measuring techniques

The test fixture was selected depending on the nature of the material to be experimented with. Some significant features behind the considered elected option are the frequency range, the required accuracy level, the measurement volume, and the experiments' budget [31]. Consequent to the dielectric material under test, the

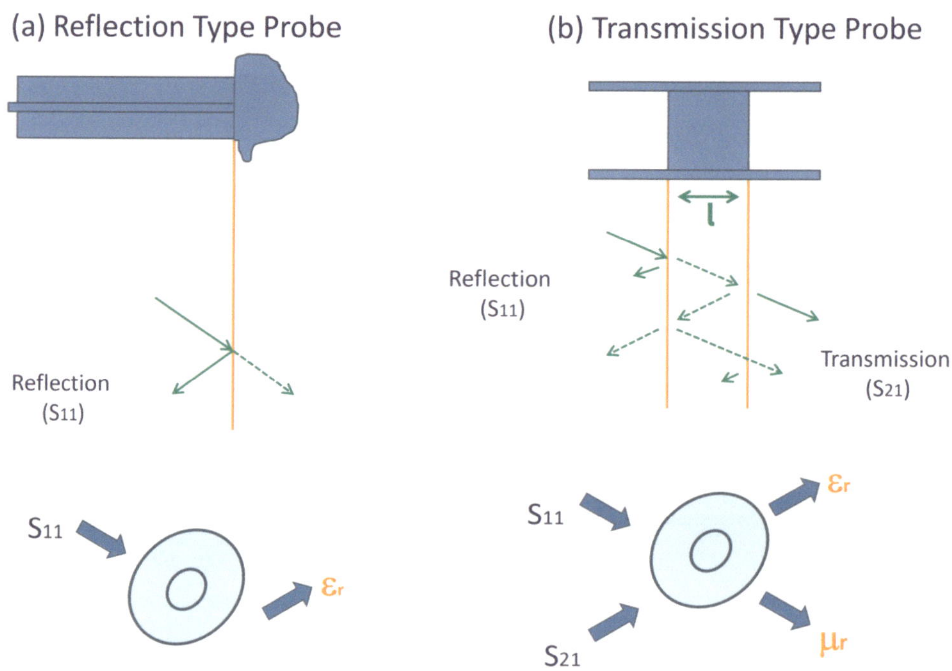

Figure 2.
Reflection and transmission types of probes [31].

design of the sample holder, the measurement system, and the technique was designated.

Figure 3 illustrates the main methodologies employed in the agricultural and food industry in relation to frequency.

The coaxial probe method is a simple and convenient method for liquids and semi-powders where the material is measured by emerging the probe into the liquid and such a system consists of the probe, the network analyzer, the software that is included in the 85070E probe kit. No external computer is needed for the PNA generations. At the same time, it is not the case for other analyzers that rely on the 82357A USB (universal serial bus) to GPIB (General Purpose Interface Bus) interface. A common technique is the Agilent 85070E open coaxial probe, which has been used immensely in biological materials, liquids, and other mixtures, where the permittivity can be derived at a frequency ranging from 0.2 to 50 GHz [31]. The vector network analyzer accompanying such experiments is the popular Agilent 87070 [31]. This system is premeditated to be highly priced but was extensively used in the industry, being one of the most commercial measurement systems adapted for solids and liquids.

Most of the food industry experiments in the last decade were executed using the 85070E Dielectric Probe Kit, including probe and software. The complete system used in process analytic technologies is based on a network analyzer measuring the dielectric parameters of the material (dielectric constant, loss factor, loss tangent, and Cole-Cole). It is connected to the Vector network analyzer Agilent N9912A.

Currently, a commercially available THz-TDS system (TeraView TPS 4000) is employed, which incorporates a pair of GaAs photoconductive antennas (emitter and receiver). This system utilizes a femtosecond laser module to generate laser pulses that are divided into two separate beams (pump and probe) using a beam splitter [30].

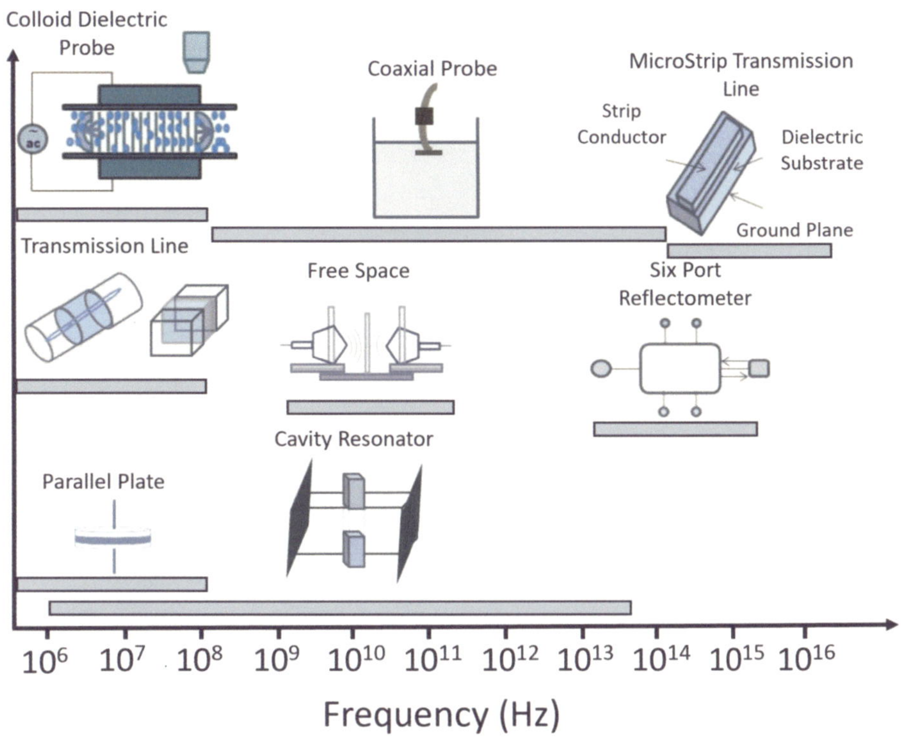

Figure 3.
Methods for characterizing dielectric properties in the agri-food sector across different frequency ranges [31].

The transmission line technique is primarily employed for machinable solids that can be inserted into the transmission line. This broadband method is limited in frequency coverage only by the size of the sample holder.

A technique that requires no contact is the free space approach. This method relies on antennas to focus microwave energy through a slab of the material under elevated temperature, mainly used at millimeter wave frequency.

Within a specific frequency range, the cavity perturbation technique has been widely employed for liquid samples at both low and high temperatures. The resonant cavities at a specified frequency are influenced by the material sample, allowing for the calculation of permittivity. Notable options include Keysight's 85072A 10 GHz split-cylinder resonator and split-post dielectric resonators [27].

The fundamental inexpensive technique widely applied is the parallel plate capacitor method, where the material is placed between the electrodes forming a capacitor. An impedance analyzer or even an LCR meter can be adopted for a typical system. Another method relies on inductance measurement to derive the permeability where the material under test is wrapped with a wire, and inductance is evaluated with respect to the ends of the wire. For this technique, the Keysight 16454A magnetic material test fixture is available. When a toroidal core is put in, this probe does not flux and thus is considered ideal for single-turn inductors. Reflection-based measurements offer a versatile approach to investigate solids and liquids over a broad frequency spectrum. Alternatively, by employing the Fourier transform of the sensor's reflectogram resulting from the applied forcing pulse, the frequency spectrum of the complex dielectric permittivity can be obtained. To construct a dielectric permittivity

DOI: http://dx.doi.org/10.5772/intechopen.1002493

sensor, two or three stainless steel rods can be strategically positioned within the material under examination, forming a segment of a parallel waveguide.

Graphical representation depicted in **Figure 4** illustrates the system's measurement suite, classified according to the material type being tested and the frequency range employed for measurements.

Some genuine dielectric properties experiments were recently witnessed at the Institute of Agrophysics in Lublin. A web server was used to store data concerning the soil moisture, temperature, and salinity of the soil. Such an implementation has some advantages in lowering power consumption in the long term [36].

While extensive scientific research on dielectric properties explores the testing of various objects, achieving accurate characterization across a wide frequency range through dielectric spectroscopy remains challenging. Therefore, improving dielectric measurements and developing advanced sensors can expand the field's investigation scope. Additionally, designing equipment for radiofrequency and MW dielectric heating applications holds significant importance [37].

3. Applications

Although it has extended to medicine and material engineering, the main field of research for dielectric characterization has been the horticultural field. The number of applications of these techniques is extensive, and they are usually classified according to the technique used, appearing in two large divisions, dielectric characterization studies and BIS measurement studies [38].

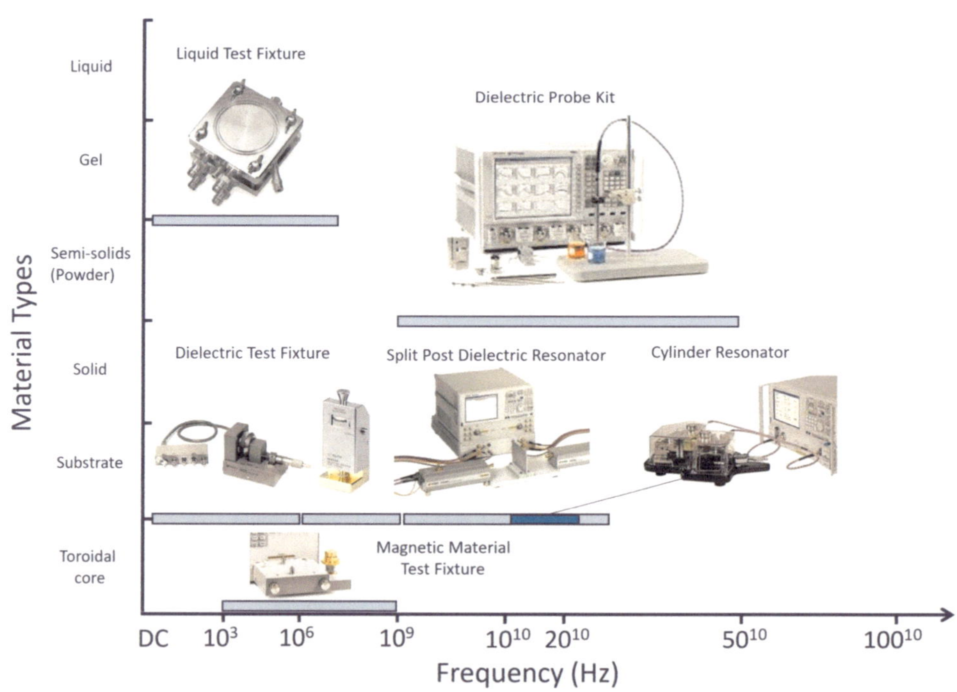

Figure 4.
Measurement apparatus used for the measurement of materials [12].

The quality factors of fruits and vegetables that are notably worth highlighting are moisture content, maturity defects, and imperfections. All these factors help harvest, classify, and package the products, improving their uniformity and quality [39]. Conducting this type of measurement using simple and innocuous techniques has been a subject of constant development in recent years, with dielectric characterization being a handy tool in food engineering and technology. Initially, the quality studies contemplated the properties of fruit and vegetables [40]. Subsequently, they have spread to other tissues of animal origin, such as meat [41] or fish [42].

The dielectric properties correlate with the quality parameters of the objects, as well as with their physical characteristics and chemical properties [12]. Quality parameters encompass firmness, pH, soluble solids, moisture content, and electrical conductivity. It is essential to ascertain the ideal frequency range that yields the desired dielectric properties, the depth of penetration reached in a particular food, and its variability with frequency or temperature to characterize varied materials.

Table 1 summarizes the main advances established in studies and applications of dielectric characterization in recent years.

Electrical impedance spectroscopy is a method employed to evaluate the electrical characteristics of substances through the application of alternating electrical signals at various frequencies, followed by the measurement of the resulting response signals. This technique is safe, non-invasive, fast, portable, low cost, and easy to use, thereby holding significant potential for monitoring quality processes within the food industry [107].

The structure and composition of the materials determine the dielectric properties, while the frequency and temperature are related to the maturity of the agricultural product. Furthermore, ionic components exert notable repercussions on dielectric properties. Density is one more crucial element affecting the interaction between the electromagnetic field and the mass involved. The storage time of agricultural products also plays a vital role in their dielectric properties, as maturation processes occurring during storage can affect them.

The focal point of this investigation involves examining the interrelationships between the dielectric properties and additional chemical and physical attributes of the analyzed substance. The utilization of rapid and non-destructive quality assessment techniques for agricultural products is of utmost importance when evaluating dielectric spectroscopy methodologies in the field of agrophysics [9]. These techniques offer a clean and non-destructive approach that is imperative for effective management systems. They can provide a means to simultaneously examine changes in fruit tissue during ripening and intracellular and extracellular behaviors. Impedance spectroscopy has succeeded in a very high-frequency range, reaching up to 6.25 MHz.

Among the main BIS applications, it is worth highlighting the determination of freshness in fruits of different kinds, such as apples [108] and bananas [109]. As well as the measure of aging [110] or the ripening of apples [111], mangos [112], citrus [113] and pears [114], and other fruits. This technique can also build an immunosensor that looks for virus selection in products [115] and discriminates between healthy and infected samples [116]. Determining acidity in citrus using dielectric impedance has been particularly interesting in recent years as it is a non-destructive method [117]. The effects of temperature on kakis [118], nectarines [119], and their content of soluble solids [120] have been established.

BIS can monitor the concentration of elements such as nitrogen in plants [121]. Therefore, it is considered a non-destructive technique for estimating the impacts of

Food	Research	Reference
 Apples	Relationship with quality during ripening on the tree	[40, 43]
	Maturation and shelf life	[44]
	Identification of the varieties	[45]
	Bruised product quality	[46]
	Quality parameters	[9, 46]
	Maturity parameters	[47]
	Determination of soluble solids content	[48]
	Quality factors during 10 weeks of storage	[43]
	Decreases with temperature and frequency	[48]
 Apple juice	Temperature dependence similar to water	[4]
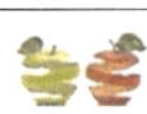 Apple peel	The dielectric constant decreases with temperature	[40]
 Carrots	Variation with frequency, enormous values at low frequencies	[49]
	Inflection points and critical edge frequency at 100 MHz	[50]
	Internal structure and physiological state of tissues	[51]
 Peas	Changes in the salt content and sample thickness	[52]
 Eggplants	Increase with increasing storage period	[52]
	Energy storage capacity and energy loss vary depending on ambient temperature	[53]
 Avocados	Temperature dependence disappears at a frequency of 100 MHz	[54]
	It decreases with increasing temperature	[55]
 Coconut water	The energy storage capacity reduces with frequency and is minimum around 2 GHz	[53]
 Coconut oil	Use as electrical insulating material	[56]
 Grapes	Loss factor increases with storage time	[12]
 Grape juice/ wine	Variation of dielectric properties with time of change of state	[57]
	Relationships with the variety or the area of the collection can influence the outcome	[58]
	Control of wine fermentation	[59]
	Dependence on humidity and temperature	[60]
	Relaxation frequencies according to the molecular structure of the varieties	[61]
	Determination of adulteration	[62]

Food	Research	Reference
	Dielectric constant and loss factor are higher in grape juice than in wine at 200 MHz	[58]
	Dielectric constant decreases with frequency	[57]
 Guavas	The depth of penetration showed a decrease as the frequency and temperature increased	[63]
	Dielectric constant decreased with temperature for frequencies above 1000 MHz	[64]
 Mangos	Physicochemical transformations occurred in the ripening process	[65]
	Sugar content, lowest value on the day of peak maturity	[66]
	Ripening characteristics	[67]
	Relationship with loss factor, critical frequency, pH, and moisture content during preharvest	[68]
	Relationship with fructose and pH in the 2.5–12.5 GHz range	[69]
	Dielectric constant and the loss factor decrease as humidity decreases at low frequencies	[70]
	Sound product properties (0.5–20 GHz)	[71]
 Melon	Linear relationship with the content of soluble solids	[72]
	Classification of varieties	[73]
	Quality parameters	[74]
 Watermelon	Detection of maturity of 10 MHz–1.8 GHz and quality parameters	[75]
	Decreased permeability with humidity	[47]
	Determination of sugar content	[76]
 Tomatoes	Maturity	[77]
	The values exhibit a downward trend as the frequency increases. Additionally, at elevated temperatures, there is a decrease in the dielectric constant and an increase in the dielectric loss	[78]
	Determination of varieties	[79]
	The depth of penetration exhibits a reduction as the frequency and loss factor increase	[80]
 Potatoes	Quality evaluation	[47]
	Dielectric properties decrease with frequency	[81]
 Pears	Determination of firmness	[82]
	Qualitative characteristics.	[83]
	Higher values for higher bruise levels and lower moisture contents	[84]
 Peaches	Values decrease with increasing frequency and temperature	[85]
	Non-linear relationship with soluble solids, moisture content, and pH	[86]
 Oranges	Evaluate penetration depth of microwaves and design applicators for pasteurization processes	[87]
 Cane sugar	Variation with sugar concentration and temperature	[88]

Food	Research	Reference
 Honey	Honey adulteration	[89]
 Nuts	Moisture content	[90]
	Values increase with temperature	[91]
 Legumes	Increase with temperature and moisture concentration, decrease with frequency	[92]
	There is an inverse linear correlation observed between the loss factor and the frequency	[93]
	Depth of penetration decreases with increasing frequency	[94]
 Milk	Decreases with increasing frequency and varies with storage	[95]
	Measurement of sugar content and relationship with fat content	[96]
	Variation of properties with frequency	[97]
 Vinegar	Dielectric constant decreases as frequency increases	[98]
 Animal products	Relationship with temperature and humidity	[99]
	Identification of sex and age of the species	[100]
	Determination of metabolic state	[41]
 Eggs	Quality parameters	[42]
	Composition	[101]
	Quality classification	[102]
	Freshness	[103]
	Decreased with frequency, less in the bud	[104]
 Liquids	Measurement of alcohol content	[105]
 Oils	Composition, adulteration and quality	[106]

Table 1.
Dielectric characterization applications and studies in recent years (1949–2023).

other ions on plant tissue, achieving higher crop yields. Additionally, BIS enables monitoring of plant conditions, including early developmental stages, and facilitates quality control of the final products.

Impedance measurement has recently been employed to investigate the isolated effect of viruses [122]. The findings reveal that the resistance of the leaves of the infected plant is lower than that of the leaves of the healthy plants. Therefore, the impedance measurement is a fast and intelligent method to diagnose the existence of diseases in plants [123], the relaxation processes [10], or the diffusion and interaction of biomolecules with water [124].

BIS is an intriguing method to assess the condition or composition of human organs and different types of biological tissues in vivo [125]. Given the significance of food disinfection for public health and safety, numerous related applications exist. An example of this is the development of an innovative methodology for the real-time detection of the amount of disinfectant in a sample by determining the characteristic frequencies and the dielectric permittivity spectra within a specific frequency range. This method establishes a correlation between the characteristic frequencies and disinfectant concentration with acceptable precision [126].

Among postharvest applications, radiofrequency (RF) and microwave (MW) heating affects the dielectric properties of materials. Permittivity and moisture content are closely correlated in food with high water content. It is essential to find the dielectric properties of partially frozen materials to assess the heating rates and uniformity during microwave (MW) thawing [51]. Water in the microwaved material can increase heating efficiency, as water is a strong microwave absorber that converts microwave energy into heat. However, when the moisture content is too high, there can be a negative effect on heating efficiency, as the water can absorb so much energy that thermal overload occurs in the material, leading to degradation or even carbonization. Therefore, the relationship between moisture content and microwave heating efficiency is intricate and influenced by many factors, such as the composition and structure of the material, the frequency and intensity of microwaves, and the duration of heating. These properties are essential to detect processing conditions or food quality [127].

Dielectric heating has shown its applicability for fruit drying, often achieving quickly and providing higher quality dried products with low energy consumption [128]. Through RF and dielectric heating applied in the drying of agricultural products, it improves thermal efficiency, the quality of the final product and reduces drying time compared to conventional drying. Using RF and dielectric heating, food can be protected from insects inside the nuts [129]. The dielectric heating methodology can be used to reduce the necessary times of heat treatment that entail the control of pests present in food after harvest [130].

The pasteurization and sterilization of fruit juice concentrate are processes where the proper choice of frequency is essential to obtain a uniform pasteurization [131].

In this context, considering the significance of the characterization process of fruits and vegetables from harvest to cold storage, along with the accelerated technological advancements and the availability of mathematical methods, as well as the numerous research studies conducted worldwide, there is a significant demand to review the existing electrical characterization methods in the horticultural field.

4. Conclusions

The robust association between the dielectric response and the chemical and physical composition of agro-physical materials enables the identification of specific properties and quality indicators related to them. One of the primary applications of DS and BIS is the assessment of fruit ripeness and quality. As fresh produce ripens, its dielectric properties change, utilized as indicators of ripeness and quality. The detection of dielectric properties of fresh fruit can also be used to detect disease and mechanical damage. These techniques can also be used to determine the quality of the products during storage, and therefore their control prevents loss of quality and freshness. In addition, they can be used to evaluate product quality during processing to meet quality and freshness standards since techniques such as pasteurization or freezing affect their dielectric properties.

The development of particular and advanced techniques and instrumentation improves the capabilities of these non-destructive characterization techniques, as they are valuable for environmental and economic scientific evaluation within the food industry. Accuracy is indispensable in the measurement of dielectric and bioelectrical impedance spectroscopy since some essential characteristics must be maintained during the execution of the measurement. Factors such as the probe's altitude, the material's manufacture, and the quality and size of the beaker used must be considered in the experimentation and measurement. In addition, the amount of liquid under the test in which the probe is fused, the agreement with the amount of water under which the calibration has been performed, and an essential parameter such as the constant depth of the measurement must be considered. Advancements in understanding the dielectric characteristics of materials have led to further progress in the sensing field, enabling the monitoring of fresh produce's properties. The objective is to maintain the optimal conditions of fresh produce from production and harvesting to consumption.

Despite the antiquity of applying these non-destructive techniques, they are still experimental techniques under development, with a solid technological impact and providing numerous novelties. The traditional field of research has been the fruit and vegetable sector, although nowadays, the study has been extended to different areas, thus establishing new challenges and future perspectives.

Author details

Nuria Novas Castellano[1*], Dalia El Khaled[1,2], Rosa Maria Garcia[1], Francisco Portillo[1], Manuel Fernadez-Ros[1] and Jose Antonio Gazquez[1]

1 Research Centre in Intensive Mediterranean Agrosystems and Agri-Food Biotechnology of the University of Almeria (CIAIMBITAL), University of Almería, Almería, Spain

2 Arab East Colleges, Al Qirawan, Riyahd, Saudi Arabia

*Address all correspondence to: nnovas@ual.es

References

[1] Novas N et al. Development of a smartphone application for assessment of chilling injuries in zucchini. Biosystems Engineering. 2019;**181**:114-127

[2] Xu Y, Li C, Wang J. How does agricultural global value chain affect ecological footprint? The moderating role of environmental regulation. Sustainable Development. 2023; (February):1-12

[3] Palumbo M et al. Emerging postharvest technologies to enhance the shelf-life of fruit and vegetables: An overview. Food. 2022;**11**(23):1-29

[4] Al Faruq A et al. New understandings of how dielectric properties of fruits and vegetables are affected by heat-induced dehydration: A review. Drying Technology. 2019;**37**(14):1780-1792. DOI: 10.1080/07373937.2018.1538157

[5] Mahanti NK et al. Emerging non-destructive imaging techniques for fruit damage detection: Image processing and analysis. Trends in Food Science and Technology. 2022;**120**(October 2021): 418-438. DOI: 10.1016/j.tifs.2021.12.021

[6] Guo B et al. Bruise detection and classification of strawberries based on thermal images. Food and Bioprocess Technology. 2022;**15**(5):1133-1141. DOI: 10.1007/s11947-022-02804-5

[7] Cavaco M et al. Nondestructive assessment of citrus fruit quality and ripening by visible–near infrared reflectance spectroscopy. In: Muhammad Sarwar Khan M, Ahmad Khan I, editor. Citrus—Research, Development and Biotechnology. Chapter 13. UK: IntechOpen; 2021. pp. 1-30

[8] Zhu D et al. Collaborative analysis on difference of apple fruits flavour using electronic nose and electronic tongue. Scientia Horticulturae (Amsterdam). 2020;**260**(August 2019):108879. DOI: 10.1016/j.scienta.2019.108879

[9] Ali MM, Hashim N. Non-destructive methods for detection of food quality [Internet]. In: Future Foods: Global Trends, Opportunities, and Sustainability Challenges. Amsterdam, Netherlands: Elsevier Inc.; 2021. pp. 645-667. DOI: 10.1016/B978-0-323-91001-9.00003-7

[10] Li Y et al. Radio-frequency dielectric relaxation behavior of selected vegetable tissues: Spectra analysis with logarithmic derivative method and simulation with double-shell model. Journal of Food Engineering. 2020;**277**(September 2019): 109914. DOI: 10.1016/j.jfoodeng.2020.109914

[11] El Khaled D et al. Cleaner quality control system using bioimpedance methods: A review for fruits and vegetables. Journal of Cleaner Production. 2017. DOI: 10.1016/j.jclepro.2015.10.096

[12] El Khaled D et al. Fruit and vegetable quality assessment via dielectric sensing. Sensors (Switzerland). 2015;**15**(7):15363-15397

[13] Kongshuang Z, Yuan L. A review of the application of dielectric spectroscopy in food field. Food Science. 2019; **40**(19):0-1

[14] You KY, Sim MS, Abdullah SN. Emerging microwave technologies for agricultural and food processing. Precision Agriculture Technologies for Food Security and Sustainability. Chapter 5. Hershey, Pennsylvania, USA: IGI Global; 2020. pp. 94-148

[15] Nelson SO et al. Assessment of microwave permittivity for sensing peach maturity. Transactions of ASAE. 1995;**38**(2):579-585

[16] Tan DQ. The search for enhanced dielectric strength of polymer-based dielectrics: A focused review on polymer nanocomposites. Journal of Applied Polymer Science. 2020;**137**(33):1-32

[17] Marzec A et al. Structural, optical and electrical properties of nanocrystalline TiO_2, SnO_2 and their composites obtained by the sol-gel method. Journal of the European Ceramic Society. 2016;**36**(12):2981-2989

[18] Anatolii S et al. Model of combined solid plasma material for the protection of radio-electronic means of optical and radio radiation. International Journal of Advanced Trends in Computer Science and Engineering. 2019;**8**(4): 1241-1247

[19] González M, Pozuelo J, Baselga J. Electromagnetic shielding materials in GHz range. Chemical Record. 2018;**18**(7):1000-1009

[20] Szychta L et al. The dielectric properties of worker bee homogenate in a high frequency electric field. Energies. 2022;**15**(24): 9342

[21] Ravi Kumar K, Krishna Chaitanya NVV, Sendhil KN. Solar thermal energy technologies and its applications for process heating and power generation—A review. Journal of Cleaner Production. 2021;**282**:125296

[22] Zhao B et al. Achieving wideband microwave absorption properties in PVDF nanocomposite foams with an ultra-low MWCNT content by introducing a microcellular structure. Journal of Materials Chemistry C. 2019;**8**(1):58-70

[23] Munawar T et al. Multi metal oxide NiO-CdO-ZnO nanocomposite–synthesis, structural, optical, electrical properties and enhanced sunlight driven photocatalytic activity. Ceramics International. 2020;**46**(2):2421-2437. DOI: 10.1016/j.ceramint.2019.09.236

[24] El Khaled D et al. Microwave dielectric heating: Applications on metals processing. Renewable and Sustainable Energy Reviews. 2018;**82**(November 2017):2880-2892. DOI: 10.1016/j.rser.2017.10.043

[25] Prateek TVK, Gupta RK. Recent Progress on ferroelectric polymer-based nanocomposites for high energy density capacitors: Synthesis, dielectric properties, and future aspects. Chemical Reviews. 2016;**116**(7):4260-4317

[26] Meng Z, Wu Z, Gray J. Microwave sensor technologies for food evaluation and analysis: Methods, challenges and solutions. Transactions of the Institute of Measurement and Control. 2018;**40**(12): 3433-3448

[27] Keysight. 2023. pp. 1-6. Available from: https://www.keysight.com/us/en/home.html

[28] Ahmed OF, Thaher RH, Ahmed SR. Design and fabrication of UWB microstrip Antenna on different substrates for wireless Communication system. In: 2022 - International Congress on Human-Computer Interaction, Optimization and Robotic Applications (HORA). Vol. 3(7). Ankara, Turkey. 2022. pp. 7-10

[29] Dunster JL et al. Mathematical techniques for understanding platelet regulation and the development of new pharmacological approaches. Methods in Molecular Biology. 2018;**1812**:255-279

[30] Landinger TF, Schwarzberger G, Jossen A. A novel method for high

frequency battery impedance measurements. In: 2019 IEEE International Symposium on Electromagnetic Compatibility, Signal & Power Integrity (EMC+SIPI), New Orleans, LA, USA. 2019. pp. 106-110

[31] El Khaled D et al. Dielectric spectroscopy in biomaterials: Agrophysics. Materials (Basel). 2016;**9**(5):310

[32] Skierucha W, Walczak R, Wilczek A. Comparison of open-ended coax and TDR sensors for the measurement of soil dielectric permittivity in microwave frequencies. International Agrophysics. 2004;**18**(4):355-362

[33] Oliveira JGD, Junior JGD, Pinto ENMG, Neto VPS, D'Assunção AG. A new planar microwave sensor for building materials complex permittivity characterization. Sensors (Switzerland). 2020;**20**(21):1-15

[34] He H et al. Topological negative refraction of surface acoustic waves in a Weyl phononic crystal. Nature. 2018; **560**(7716):61-64. DOI: 10.1038/s41586-018-0367-9

[35] Kopetz H. Real-time systems: Design principles for distributed embedded applications. In: Real-Time Systems Series. Vol. 34. New York, NY: Springer; 1997. 142 p

[36] Park C et al. A dielectric mixing model accounting for soil organic matter. Vadose Zone Journal. 2019;**18**(1):190036

[37] Macana RJ, Baik OD. Disinfestation of insect pests in stored agricultural materials using microwave and radio frequency heating: A review. Food Review International. 2018;**34**(5):483-510

[38] Qu S et al. Dielectric and magnetic loss behavior of nanooxides. In: Spectroscopic Methods for Nanomaterials Characterization. Netherlands: Elsevier Science; 2017. pp. 301-319

[39] Ragni L et al. Quality evaluation of shell eggs during storage using a dielectric technique. Transactions of the ASABE. 2007;**50**(4):1331-1340

[40] Guo W et al. Maturity effects on dielectric properties of apples from 10 to 4500 MHz. LWT—Food Science and Technology. 2011;**44**(1):224-230. DOI: 10.1016/j.lwt.2010.05.032

[41] Traffano-Schiffo MV et al. New methodology to analyze the dielectric properties in radiofrequency and microwave ranges in chicken meat during postmortem time. Journal of Food Engineering. 2021;**292**(September 2020):110350. DOI: 10.1016/j.jfoodeng.2020.110350

[42] He J, Li F, Jiao Y. Prediction of salmon (*Salmo salar*) quality during refrigeration storage based on dielectric properties. International Journal of Agricultural and Biological Engineering. 2021;**14**(4):262-269

[43] De Vasconcelos D, Andrade Pina VA. Postharvest quality parameters evolution in 'Golden delicious', 'Gala' and 'Starking' apple. KnE Engineering. 2020;**5**(6):51-62

[44] Kafarski M et al. Evaluation of apple maturity with two types of dielectric probes. Sensors (Switzerland). 2018;**18**(1):121

[45] Shang L, Guo W, Nelson SO. Apple variety identification based on dielectric spectra and chemometric methods. Food Analytical Methods. 2015;**8**(4):1042-1052

[46] Bian HX et al. Quality predictions for bruised apples based on dielectric

properties. Journal of Food Processing & Preservation. 2019;**43**(8):1-10

[47] Ikediala JN et al. Dielectric properties of apple cultivars and codling moth larvae. Transactions of the American Society of Agricultural Engineers. 2000;**43**(5):1175-1184

[48] Solyom K et al. Effect of temperature and moisture contents on dielectric properties at 2.45 GHz of fruit and vegetable processing by-products. RSC Advances. 2020;**10**(28):16783-16790

[49] Schwan HP. Electrical properties of tissue and cell suspensions: Mechanisms and models. In: Proceedings of 16th Annual International Conference of the IEEE Engineering in Medicine and Biology Society, Baltimore, MD, USA, 1994. pp. A70-A71

[50] Shaw TM, Galvin JA. High-frequency-heating characteristics of vegetable tissues determined from electrical-conductivity measurements. Proceedings of the IRE. 1949;**37**(1):83-86

[51] Zhang C et al. In-situ dielectric analysis on the responses of vegetable cells to freezing treatment. International Journal of Refrigeration. 2020;**118**:384-391. DOI: 10.1016/j.ijrefrig.2020.06.029

[52] Jain D et al. Effect of changes in salt content and food thickness on electromagnetic heating of rice, mashed potatoes and peas in 915 MHz single mode microwave cavity. Food Research International. 2019;**119**(January 2018): 584-595. DOI: 10.1016/j.foodres.2018.10.036

[53] Kundu A, Gupta B. Broadband dielectric properties measurement of some vegetables and fruits using open ended coaxial probe technique. Int Conf Control Instrumentation, Energy Commun CIEC. 1980;**2014**(2014): 480-484

[54] Nelson SO. Frequency and Temperature permittivities of fresh fruits and vegetables from 0.01 to 1.8 GHz. Trans Agric Res Serv. 2018;**46**(2): 567–574

[55] Coronel P et al. Dielectric properties of pumpable food materials at 915 MHz. International Journal of Food Properties. 2008;**11**(3):508-518

[56] Pambudi NA, Yusuf AM, Sarifudin A. The use of single-phase immersion cooling by using two types of dielectric fluid for data Center energy savings. Energy Eng J Assoc Energy Eng. 2022; **119**(1):275-286

[57] García A et al. Dielectric characteristics of grape juice and wine. Biosystems Engineering. 2004;**88**(3): 343-349

[58] García A et al. Dielectric properties of grape juice at 0.2 and 3 GHz. Journal of Food Engineering. 2001;**48**(3):203-211

[59] Zheng S, Fang Q, Cosic I. An investigation on dielectric properties of major constituents of grape must using electrochemical impedance spectroscopy. European Food Research and Technology. 2009;**229**(6):887-897

[60] Sólyom K et al. Dielectric properties of grape marc: Effect of temperature, moisture content and sample preparation method. Journal of Food Engineering. 2013;**119**(1):33-39

[61] Vijay R, Jain R, Sharma KS. Dielectric spectroscopy of grape juice at microwave frequencies. International Agrophysics. 2015;**29**(2):239-246

[62] Naderi-Boldaji M et al. Potential of two dielectric spectroscopy techniques

and chemometric analyses for detection of adulteration in grape syrup. Journal of the International Measurement Confederation. 2018;**127**(March):518-524. DOI: 10.1016/j.measurement.2018.06.015

[63] Kataria TK et al. Dielectric properties of guava, mamey sapote, prickly pears, and nopal in the microwave range. International Journal of Food Properties. 2017;**20**(12):2944-2953. DOI: 10.1080/10942912.2016.1261154

[64] González-Monroy AD et al. Dielectric properties of beverages (tamarind and green) relevant to microwave-assisted pasteurization. Journal of Food Science. 2018;**83**(9):2317-2323

[65] Silva PF et al. Characterization of the dielectric properties of the Tommy Atkins mango. Journal of Microwaves, Optoelectronics and Electromagnetic Applications. 2020;**19**(1):86-93

[66] Lima CMC et al. Electromagnetic characterization of the tommy atkins mango in the maturation period. 2019 SBMO/IEEE MTT-S Int Microw Optoelectron Conf IMOC 2019. 2019; **2019**(3):2019-2021

[67] Yahaya NZ et al. Microwave dielectric properties for detection of "Harumanis" mangoes ripeness. Journal of Physics Conference Series. 2018;**1083**(1):012020

[68] Suhaime N et al. Microwave technique for moisture content and pH determination during pre-harvest of mango cv. Chok anan. Sains Malaysiana. 2018;**47**(7):1571-1578

[69] Krairiksh M, Mearnchu A, Phongcharoenpanich C. Nondestructive measurement for mango inspection. In: IEEE International Symposium on Communications and Information Technology. 2004;**2004**(2):646-649

[70] Cheng EM et al. Microwave reflection based dielectric spectroscopy for moisture content in Melele mango fruit (*Mangifera indica* L.). Journal of Telecommunication, Electronic and Computer Engineering. 2018;**10**(1–14):1-6

[71] Shivamurthy HT, Matacena I, Spirito M. Dielectric measurements of mangoes from 0.5GHz to 20GHz using a custom open-ended coaxial probe. In: 2017 47th European Microwave Conference (EuMC), Nuremberg, Germany. Vol. 2017. 2017. pp. 958-961

[72] Liu D et al. Non-destructive sugar content assessment of multiple cultivars of melons by dielectric properties. Journal of the Science of Food and Agriculture. 2021;**101**(10):4308-4314

[73] Zhuanwei W et al. Non-destructive testing of melon varieties based on dielectric spectrum technology. Journal of Agricultural Engineering. 2017;**33**(9):290-295

[74] Nelson SO, Trabelsi S. Examination of dielectric spectroscopy data for correlations with melon quality. Am Soc Agric Biol Eng Annu Int Meet 2011, ASABE 2011. 2011;**6**(July 2018):5123-5128

[75] Nelson SO, et al. Dielectric spectroscopy studies on honeydew melons. 2007 ASABE Annu Int Meet Tech Pap. 2007;7 BOOK(07)

[76] Isa MM et al. Sugar content in watermelon juice based on dielectric properties at 10.45GHz. In: 2009 IEEE Student Conference on Research and Development (SCOReD), Serdang, Malaysia. 2009. pp. 529-532

[77] Li J et al. Maturity assessment of tomato fruit based on electrical impedance spectroscopy. International Journal of Agricultural and Biological Engineering. 2019;**12**(4):154-161

[78] Lurwan MM et al. Dielectric properties of fresh Roma and Cherry tomato samples at different frequencies and temperatures. Journal of Science Education and Technology. 2021;**9**(4): 2021

[79] Türker U, Talebpour B, Yegül U. Determination of the relationship between apparent soil electrical conductivity with pomological properties and yield in different apple varieties. Zemdirbyste. 2011;**98**(3):307-314

[80] Abea A et al. Combined effect of temperature and oil and salt contents on the variation of dielectric properties of a tomato-based homogenate. Food. 2021; **10**(12):3124

[81] Luo GY et al. Optimization of the microwave drying process for potato chips based on the measurement of dielectric properties. Drying Technology. 2019;**37**(11):1329-1339. DOI: 10.1080/07373937.2018.1500482

[82] Zhang H et al. Non-destructive detection of the fruit firmness of Korla fragrant pear based on electrical properties. International Journal of Agricultural and Biological Engineering. 2022;**15**(6):216-221

[83] Mahmoodi MJ, Azadbakht M. Investigating the effects of qualitative properties on pears dielectric coefficient. Journal of Agricultural Machinery. 2021; **11**(1):71-81. Available from: https://jame.um.ac.ir/article_34585.html?lang=en

[84] Azadbakht M et al. Relationship of pears' dielectric properties and rates of pears' bruise. Agricultural Engineering International: CIGR Journal. 2020;**22**(1): 169-179

[85] Lombardo R, Rubino T, Cammalleri M. Dielectric characterization of fruit nectars at low RF frequencies. International Journal of Food Properties. 2015;**18**(10):2312-2326

[86] Zhang G et al. A comprehensive peach fruit quality evaluation method for grading and consumption. Applied Sciences. 2020;**10**:1348

[87] Franco AP, Tadini CC, Wilhelms Gut JA. Predicting the dielectric behavior of orange and other citrus fruit juices at 915 and 2450 MHz. International Journal of Food Properties. 2017;**20**(2):1468-1488

[88] Sumranbumrung R et al. Characterization model of dielectric properties of cane sugar solution over 0.5–14 GHz. IEEE Transactions on Instrumentation and Measurement. 2021;**70**:8003908

[89] Liu W, rt al. Quantitative determination of acacia honey adulteration by terahertz-frequency dielectric properties as an alternative technique. Spectrochimica Acta Part A: Molecular and Biomolecular Spectroscopy. 2022;**274**:121106

[90] Solar M, Solar A. Non-destructive determination of moisture content in hazelnut (*Corylus avellana* L.). Computers and Electronics in Agriculture. 2016;**121**:320-330

[91] Ling B et al. Dielectric properties of pistachio kernels as influenced by frequency, temperature, moisture and salt content. Food and Bioprocess Technology. 2015;**8**(2):420-430

[92] Oke AB, Baik OD. Role of moisture content, temperature, and frequency on

dielectric behaviour of red lentil and Kabuli chickpea in relation to radio frequency heating. Applied Food Research. 2022;**2**(1):100046

[93] Taheri S et al. Dielectric properties of chickpea, red and green lentil in the microwave frequency range as a function of temperature and moisture content. The Journal of Microwave Power and Electromagnetic Energy. 2018;**52**(3):198-214

[94] Cover JH. Bps 2021. In: Horabik J, editor. 20th International Workshop for Young Scientists. Lublin, Poland: BioPhys Spring; 2021. pp. 1-74

[95] Szerement J et al. The effect of storage time on dielectric properties of pasteurized milks and yoghurt. In: 2018 12th Int Conf Electromagn Wave Interact with Water Moist Subst ISEMA 2018. 2018. pp. 3-5

[96] Zhu X, Guo W, Liang Z. Determination of the fat content in cow's milk based on dielectric properties. Food and Bioprocess Technology. 2015;**8**(7): 1485-1494

[97] Nunes AC, Bohigas X, Tejada J. Dielectric study of milk for frequencies between 1 and 20 GHz. Journal of Food Engineering. 2006;**76**(2):250-255

[98] Tanaka F et al. Analysis of dielectric properties of rice vinegar and sake. Transactions of ASAE. 2002;**45**(3):733-740

[99] Sun J et al. Evaluation of fish freshness using impedance spectroscopy based on the characteristic parameter of orthogonal direction difference. Journal of the Science of Food and Agriculture. 2020;**100**(11):4124-4131

[100] Gómez-Salazar JA et al. Dielectric properties of fresh rabbit meat in the microwave range. Journal of Food Science. 2021;**86**(3):952-959

[101] Cong H et al. Dielectric properties of sea cucumbers (*Stichopus japonicus*) and model foods at 915 MHz. Journal of Food Engineering. 2012;**109**(3):635-639

[102] Li CH et al. Nondestructive detection of the gel state of preserved eggs based on dielectric impedance. Food. 2021;**10**(2):1-13

[103] Jun S et al. Non-destructive detection of egg freshness based on dielectric properties and yolk index regression model. Journal of Agricultural Engineering. 2016;**32**(21):290-295

[104] Guo W et al. Storage effects on dielectric properties of eggs from 10 to 1800 MHz. Journal of Food Science. 2007;**72**(5):E335-E340

[105] Kataria TK et al. Dielectric properties of tequila in the microwave frequency range (0.5–20 GHz) using coaxial probe. International Journal of Food Properties. 2017;**20**(1):S377-S384

[106] Salim A, Lim S. Review of recent metamaterial microfluidic sensors. Sensors (Switzerland). 2018;**18**(1):232

[107] Bauchot AD, Harker FR, Arnold WM. The use of electrical impedance spectroscopy to assess the physiological condition of kiwifruit. Postharvest Biology and Technology. 2000;**18**(1):9-18

[108] Nesheva DD et al. Effect of the sublayer thickness and furnace annealing on the crystallographic structure and grain size of nanocrystalline ZnxCd1-xSe thin films. Bulgarian Chemical Communications. 2013;**45**(B):11-17

[109] Ibba P et al. Bio-impedance and circuit parameters: An analysis for

tracking fruit ripening. Postharvest Biology and Technology. 2020;**159** (September 2019):110978

[110] Watanabe T et al. Electrical impedance estimation for apple fruit tissues during storage using Cole–Cole plots. Journal of Food Engineering. 2018; **221**:29-34

[111] Chowdhury A et al. Studying the electrical impedance variations in banana ripening using electrical impedance spectroscopy (EIS). In: Proceedings of the 2015 Third International Conference on Computer, Communication, Control and Information Technology (C3IT), Hooghly, India. 2015. pp. 21-24

[112] Figueiredo Neto A et al. Determination of mango ripening degree by electrical impedance spectroscopy. Computers and Electronics in Agriculture. 2017;**143**(May):222-226

[113] Gupta AK et al. Emerging approaches to determine maturity of citrus fruit. Critical Reviews in Food Science and Nutrition. 2022;**62**(19): 5245-5266

[114] Zhu X, Guo W, Wu X. Frequency- and temperature-dependent dielectric properties of fruit juices associated with pasteurization by dielectric heating. Journal of Food Engineering. 2012;**109** (2):258-266

[115] Ambrico M et al. Highly sensitive and practical detection of plant viruses via electrical impedance of droplets on textured silicon-based devices. Sensors (Switzerland). 2016;**16**(11)

[116] Khan MZH et al. Ultrasensitive detection of pathogenic viruses with electrochemical biosensor: State of the art. Biosensors & Bioelectronics. 2020; **166**(July):112431

[117] Chowdhury A et al. Electrical impedance spectroscopic study of mandarin orange during ripening. Journal of Food Measurement and Characterization. 2017;**11**(4):1654-1664

[118] Arnal L, Del Rio MA. Quality of persimmon fruit cv. Rojo brillante during storage at different temperatures. Spanish Journal of Agricultural Research. 2004;**2**(2):243

[119] O'Toole MD et al. Non-contact multi-frequency magnetic induction spectroscopy system for industrial-scale bio-impedance measurement. Measurement Science and Technology. 2015;**26**(3)

[120] Shen J et al. Prediction model of soluble solid content in Lingwu Changzao jujube based on dielectric Spectrum. J Agric Eng Trans. 2016;**32**(2): 369-375

[121] Zeleňáková L et al. Determination of nitrates in lettuce (*Lactuca Sativa* var. *Capitata*) from various producers by ion-selective electrode. Journal of Hygienic Engineering and Design. 2022; **40**:19-26

[122] Khaled AY et al. Early detection of diseases in plant tissue using spectroscopy–applications and limitations. Applied Spectroscopy Reviews. 2018;**53**(1):36-64

[123] Bukhamsin AH. Impedimetric plant biosensor based on minimally invasive and flexible microneedle electrodes. In: 2020 IEEE 33rd International Conference on Micro Electro Mechanical Systems (MEMS), Vancouver, BC, Canada. 2020. pp. 307-310

[124] Zhang J, Matsuura H, Shirakashi R. A method for measuring dielectric relaxation of water by NIR spectroscopy: Applicability and application to

measurement of water diffusion coefficient. Journal of Food Process Engineering;**2023**(November 2022)

[125] Kanoun O. Advanced systems for biomedical applications. Serie Smart Sensors, Measurement and Instrumentation. Springer International Publishing. 2021;**39**:1-286

[126] Liu J et al. Real-time sensing of disinfectant by a novel dielectric methodology. Sensors and Actuators A: Physical. 2023;**357**(March):114350

[127] Meda V, Orsat V, Raghavan V. Microwave heating and the dielectric properties of foods. In: The Microwave Processing of Foods: Second Edition. United Kingdom: Woodhead Publishing; 2017. pp. 23-43

[128] Bogale Teseme W, Weldemichael Weldeselassie H. Review on the study of dielectric properties of food materials. American Journal of Engineering and Technology Management. 2020;**5**(5):76

[129] Ayad E, Roshdy O, Afifi H. Effect of microwave energy on some stored product mites. Egyptian Academic Journal of Biological Sciences, B. Zoology. 2022;**14**(2):145-151

[130] Soto-Reyes N, Rojas-Laguna E, Sosa-Morales M. Modelación del calentamiento dieléctrico (microondas y radiofrecuencia) en sistemas alimenticios modelo. Temas Sel Ing Aliment. 2012;**6**(2):19-31

[131] Nguyen TP, Songsermpong S. Microwave processing technology for food safety and quality: A review. Agriculture and Natural Resources. 2022; **56**(1):57-72

Chapter 5

Charge Carrier Kinetics Studied by Microwave Absorption

Marinus Kunst and Reinhard Schwarz

Abstract

In this work, it will be shown that the change of the microwave reflection of a semiconductor by illumination can be used to determine the photoconductivity. The reliability of the method is successfully tested by monitoring the change of the position of the sample in the microwave field compared to results calculated with a simple model. Besides, the parameters determined by the measurements agree with those for a sample with known parameters (silicon). The use of illumination with fast laser pulses makes contactless study of excess charge carrier kinetics possible. Examples are given for measurements of systems difficult to access with conventional photoconductivity measurements: an amorphous semiconductor, a-Si:H, and a semiconducting powder, TiO_2.

Keywords: microwave (photo) conductivity, charge carrier kinetics, semiconductor quality, photocatalysis, carrier lifetime

1. Introduction

The kinetics and the mobilities of charge carriers are essential for the performance of semiconductor electronic devices. These processes can be studied most appropriately by investigation of the properties of excess charge carriers. In semiconductors, it is convenient to use light to induce excess charge carriers; additionally, this yields supplementary information for light-driven devices.

In the late forties semiconductor and microwave physics got a new impact after the military development during the war. The combination of both was relatively rare until, again, military interest was awakened for the investigation of coatings for airplanes, making them invisible for RADAR. This requires materials absorbing but not reflecting microwaves.

In the seventies, a radiation chemistry group in Delft (The Netherlands) published quantitative measurements of the excess conductivity excited by short electron pulses by microwave reflection in a closed waveguide system. This research initially involved the investigation of liquids and gases or solids that can be frozen from the liquid and so perfectly fit the waveguide. The dependence of the signal on the microwave frequency was used to test and to obtain quantitative results [1, 2].

In this work, the transient absorption of excess charge carriers in semiconductors in the microwave frequency range is investigated. This is performed by illumination of a sample with short laser pulses and monitoring the relative change of the microwave power reflected by the sample upon illumination, called the Time-Resolved Microwave Conductivity (TRMC) signal. This relative change is proportional to the transient conductivity, and so gives access to excess charge kinetics. A measurement configuration is used without strong requirements for the shape of the sample.

The experiments should be performed in waveguide equipment. Other experimental realizations, e.g., coaxial, appeared to be more noisy. Signals are induced by 10 ns (FWHM) pulses of an Nd:YAG laser at 1064 nm, 532 nm, and 266 nm.

Microwaves, provided by a Gunn diode, are directed to port 1 of a circulator, leading the microwaves via port 2 to the sample. The reflected microwaves are guided via port 3 to a point contact diode for transient measurements, replaced by a power head for stationary measurements. So measurements of the change of the reflected microwave power upon (pulsed) illumination can be monitored. For quantitative measurements, the sample is clamped between a variable short circuit. (a metal plate completely reflecting the microwaves that can be displaced). The TRMC signal, i.e., the relative change of the reflected microwave power upon illumination, $\Delta P/P$, is proportional to the photo conductance $\Delta S(t)$:

$$\Delta P \,/\, \mathrm{P} = \mathrm{A}\,\Delta S = \mathrm{A}\,\Delta\sigma\,\mathrm{d} \qquad (1)$$

where A is the sensitivity constant, $\Delta\sigma$ is the photoconductivity, and d is the thickness of the sample.

The (dark) conductivity can be measured by measuring the reflection of the sample using the power head for detection and numerical evaluation.

In both cases, the position of the sample in the microwave field can be changed by shifting the position of the short circuit. This is very useful to test the reliability of the measurements and of the evaluation.

The equipment presented here will be described as a closed waveguide system, with the sample exactly fitting the waveguide. So, the theoretical results must be compared to the results of the present equipment to detect the possible influence of the failing of the waveguide at the position of the sample.

2. Theory

2.1 Microwaves

The theory follows the theoretical lines of the Delft group with some changes because light excitation is used, and the samples are thin semiconductors [1–3]. The theory is developed for a closed system where the sample perfectly fits the waveguide. The field reflection coefficient is calculated starting at the short circuit (yielding the boundary condition) up to the space between sample and circulator using the evolution of the field reflection coefficient in a homogeneous medium and at the interface between two media with different optical constants to obtain the reflection coefficient between sample and the circulator. The power reflection coefficient R is given by the conjugate product of the field reflection coefficient.

The change of the power reflection coefficient under illumination of the sample $\Delta R(t)/R$ is the measurement (Time Resolved Microwave Conductivity) signal. The variable short circuit enables the displacement of the position of the sample in the microwave field. The analytical solution is intricate, and it is convenient to use numerical solutions. An example of such a simulation is given in **Figure 1**.

Figure 1a shows the power reflection coefficient R as a function of the distance between the sample and the short circuit, l. The simulation has been performed for a 1 μm sample with conductivity $\sigma = 1(\Omega cm)^{-1}$, the dielectric constant $\varepsilon = 1$ at 10GHz. By the change of the distance l between the sample and the short circuit, the position of the sample in the sinusoidal microwave field changes: the sample is in a very small (effectively zero) microwave field if the short circuit is against the sample (at a distance zero) leading to a very small microwave absorption and a reflection coefficient unity. Displacing the short circuit from the sample leads to a decrease of the reflection coefficient up to 1 cm distance where the reflection coefficient is at a minimum, the sample is at a maximum of the microwave field, and the absorption is at maximum. At larger distances, R is the reflection coefficient increases again up to l = 2 cm. This pattern repeats, and the dependence can be characterized as the absolute value of a sinus with a period of 2 cm. This is half of the microwave wavelength at 10GHz in a waveguide. Because the conductivity is detected independently of the sign of the field, this pattern represents the microwave field.

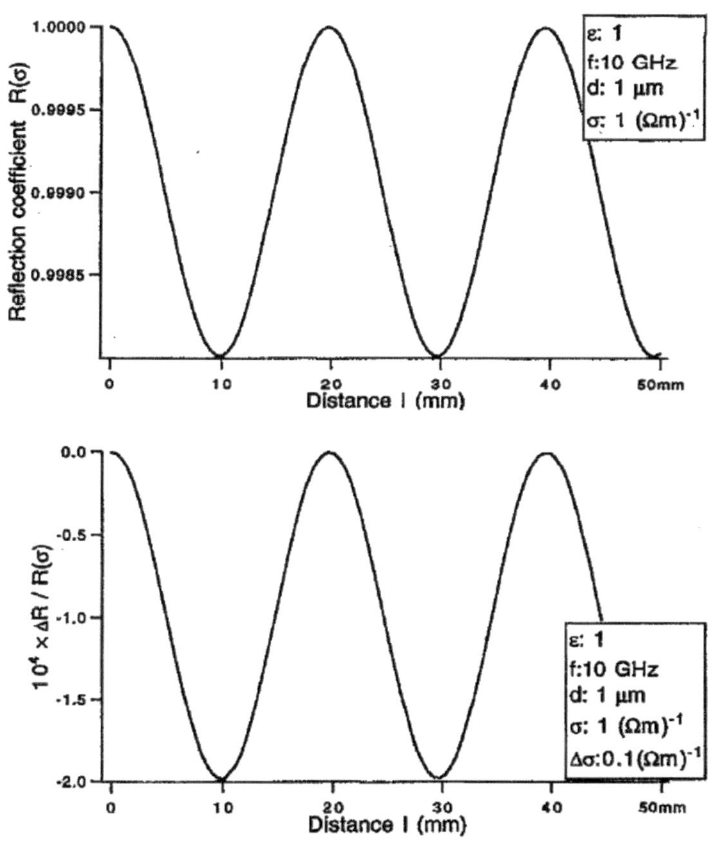

Figure 1.
Reflection coefficient R (upper curve) and ΔR/R (lower curve) in a 1-μm-thick semiconductor calculated with the model, as a function of the distance l between the short circuit and the sample (courtesy by J. Appl. Phys.).

In **Figure 1b**, the corresponding TRMC signals $\Delta R/R$ (or $\Delta P/P$) are plotted for a conductivity change $\Delta\sigma = 0.1(\Omega cm)^{-1}$. The signal is negative (increased absorption). The highest (absolute) value of the signal is found at the minimal value of R (largest absorption) and the lowest (absolute) value at the maximum value of R (very small absorption).

2.2 (Excess) charge carrier kinetics

In Si, the (excess) conductivity is due to electrons and holes in the respective bands:

$$\Delta\sigma(t) = \Delta n(t)\mu_n e + \Delta p(t)\mu_p e \qquad (2)$$

with $\Delta n(t)$ ($\Delta p(t)$) the density and μ_n (μ_p) the mobility of excess electrons (holes), respectively, and e the unity charge.

In general, the decay time τ of the signal is given by:

$$1/\tau = 1/\tau_s(t) + 1/\tau_v \qquad (3)$$

where τ_s reflects decay by surface recombination and τ_v that due to volume recombination.

Kinetics are described by the continuity equation for electrons and that for holes connected by the Poisson equation. In crystalline silicon in most cases, the densities of electrons and holes can be considered to be equal. Then, the equations can be reduced to the ambipolar continuity equation with one diffusion constant only: the ambipolar diffusion constant. This reduces to the minority carrier diffusion constant for low injection, as it is the case for the data presented below. Detailed descriptions can be found in the literature [4].

For the present work, only some particular cases are presented. Interesting is the case of a surface that can be considered as an infinite sink for charge carriers. This can easily be tested in a silicon sample by generating a signal with 532 nm light (surface excitation). Then, the decay behavior due to recombination at the illuminated surface before an appreciable number of excess charge carriers attain the other surface is:

$$\Delta n(t) \sim t^{-0.5} \qquad (4)$$

This extended decay is explained by the diffusion of excess charge carriers away from the recombinative surface [5].

For both surfaces abrased after uniform (1064 nm) excitation, the decay time after the establishment of a stationary (principal mode) distribution is [5]:

$$\tau_s = d^2 / (D\pi^2) \qquad (5)$$

where D and d are the carrier diffusivity and film thickness, respectively.

For only one surface abrased, this is given by:

$$\tau_s = 4\,d^2 / \left(D\pi^2\right) \tag{6}$$

3. Experimental section

3.1 Measurements of silicon

3.1.1 (Dark) conductivity measurements

In **Figure 2**, measurements of the reflection coefficient R of a 525 μm Si wafer as a function of l are plotted (points) and compared to the best fit of the simulation with only one freely varying parameter: the conductivity σ. For this best fit, an equivalent value 47Ωcm was found to agree with the value 40-60Ωcm specified by the producer. Also, the shape of the experimental data proves that the model explains nicely the data. So, the equipment of **Figure 3** can be described as a closed waveguide system as it is confirmed also by other measurements.

Compared to the simulated data in **Figure 1**, the data in **Figure 2** do not show a symmetrical shape. The presence of a material of a thickness that cannot be neglected with dielectric constant ε = 12 leads to deformation, although the signal is still

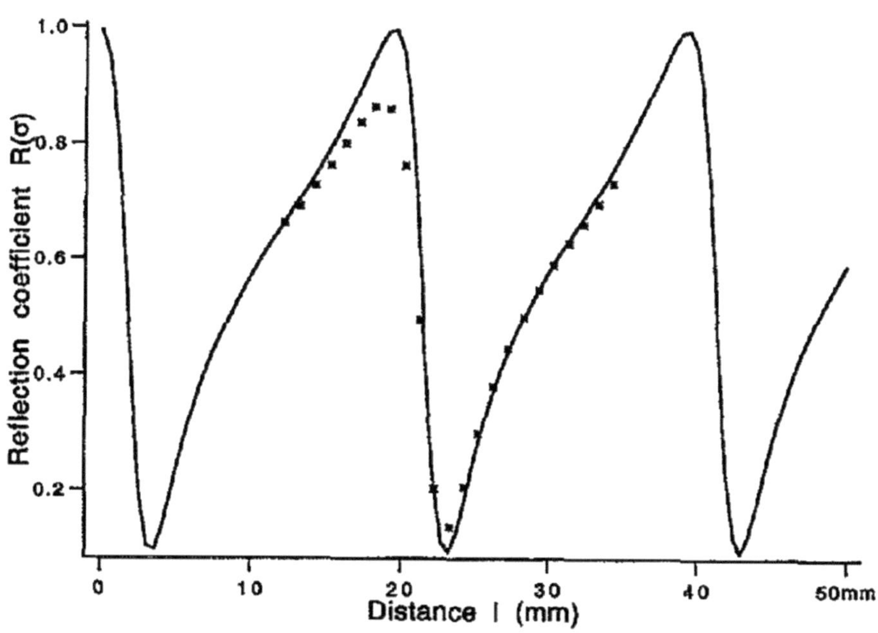

Figure 2.
Values of the reflection coefficient R measured in a 525-μm-thick n-Si wafer (40–65 Ωcm) as a function of the distance l between short circuit and the wafer (data points). The line drawn represents the best fit of the model to experimental data for σ = 2.15 (Ωcm)$^{-1}$ (i.e., ρ = 47 Ωcm). (courtesy by J. Appl. Phys.).

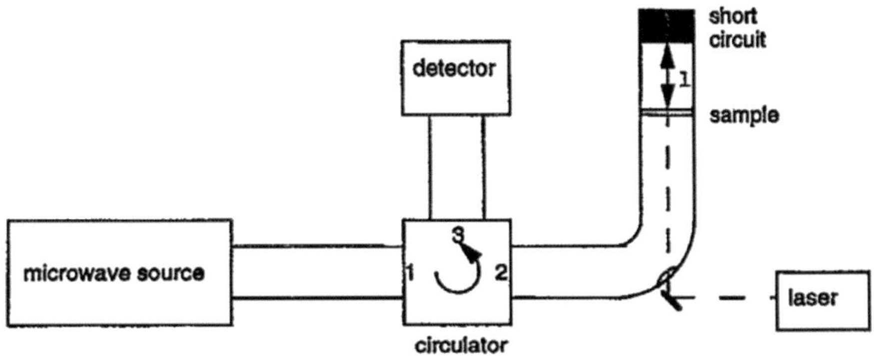

Figure 3.
Equipment for quantitative TRMC measurement.

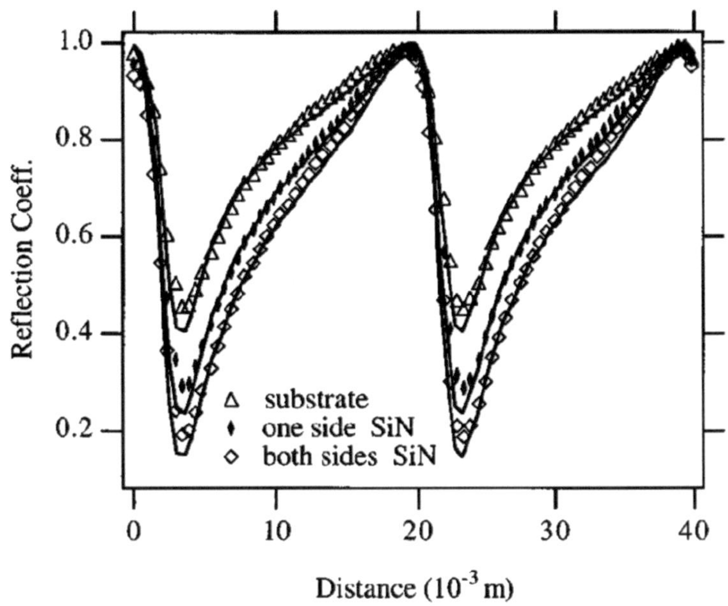

Figure 4.
The reflection coefficient R as a function of the distance l between sample and short circuit for a n-Si wafer (doping 10^{13} cm^{-3}) bare, with a 100 nm SiN film on one face and a 100 nm SiN film on both faces (courtesy by Appl. Phys. Lett.).

periodical with 2 cm. This deformation is mainly due to the decrease of the microwave wavelength in Si.

3.1.2 An application

In **Figure 4**, the power reflection coefficient R as a function of the distance l between sample and short circuit is compared for a substrate p-Si wafer (100Ωcm, 500 μm) compared to data of this wafer covered at one face with 100 nm SiN and at both faces with SiN. The best fit of the data is given by the straight lines in the figure, with the conductivity as the only variable parameter.

The experimental data show good correspondence with the simulation. Evaluation of the conductivity from the simulation shows an increase of the conductivity by the presence of the SiN film and a two times higher conductivity with both faces SiN. This cannot be due to conductivity in the film; a bare SiN film does not show appreciable photoconductivity, and besides, the charge carrier mobility is very low. Consequently, the increase of the conductivity is due to an increase of charge carriers in Si. As it is known that a SiN film on Si is positively charged, these results indicate that SiN electrons are injected into Si, probably from interface states [6]. This leads to an accumulation region in Si at the interface.

For p-type Si, the same results have been obtained. Also here, a deposited SiN layer has a positive interface. In this high ohmic sample, this is due to an inversion layer in Si. So, in both cases, electrons are emitted from (defect) states in SiN to the Si substrate.

Interestingly, the dependence of this conductivity change on the thickness of the SiN layer. Measurements of the conductivity change as a function of the SiN thickness show that this change saturates at a thickness of about 20 nm (**Figure 5**).

This suggests that 20 nm is sufficient for the passivation of Si by SiN as the passivation of Si by SiN is mainly due to the presence of a space charge. At higher dark conductivities, the analysis of the data is more difficult because also the role of the holes must be taken into account.

3.1.3 Photoconductivity (TRMC) measurements

First, it must be realized that the technique is not appropriate for the determination of the lifetime in state-of-the-art Si wafers, as the decay of the TRMC signal will be determined by surface recombination. However, modification of the surface, e.g., by deposition of lavers, or degradation of the volume, e.g., by radiation damage, can be studied informative measurements require the uniformity of the microwave field in the sample. So, the decay is only due to the immobilization of charge carriers and not to a change of position. This is conveniently done by comparison of the signals

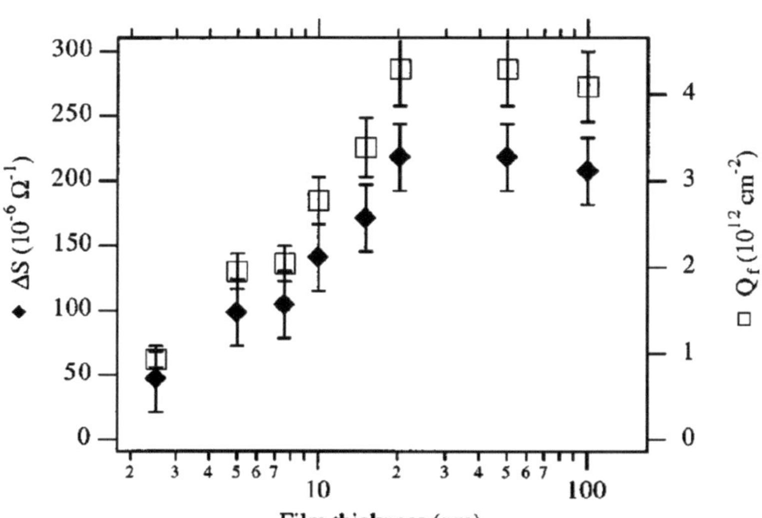

Figure 5.
Change of the conductance ΔS of a n-Si substrate as a function of the thickness of a deposited SiN film (courtesy by Appl. Phys. Lett.).

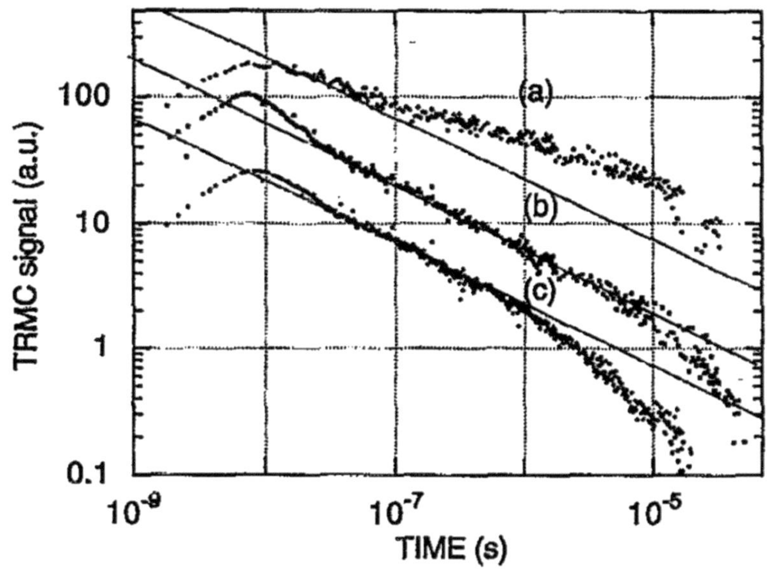

Figure 6.
TRMC transients induced by 532 nm light pulses in a 0.5 mm thick n-Si wafer (40-60Ωcm) at different distances l between the short circuit and the wafer. The label (b) refers to a position at the maximum of the microwave field, (b) and (c) to other positions. The straight lines are characterized by a slope of −0.5 (courtesy by J. Appl. Phys.).

induced by 532 nm (surface generation) to those induced by 1064 nm (uniform generation) for the same number of photons absorbed. This shows that uniformity is at best at the maximum of the microwave field [7].

An informative impression of these effects is shown in **Figure 6** for a Si wafer with abrased surfaces (i.e., infinite sinks for charge carriers):

In this figure, signals induced by 532 nm light pulses are displayed on a double logarithmic plot arbitrarily displaced in vertical direction to make the decay behavior more visible. The lines drawn have a slope 0.5: signal b was measured at the maximum of the microwave field and shows the behavior with $t^{-0.5}$ as predicted by the theory (Eq. 4). It also shows the change of the slope at the time when recombination at the other surface approximately indicated by the decay time after 1064 excitation (Eq. 5). Signal a in the figure monitored at another position in the microwave field (no uniform microwave field). Here, the decay of the signal is slower due to the diffusion of charge carriers to regions in higher microwave fields. On the contrary, signal c shows a too-fast decay. Here, the charge carriers diffuse after excitation to regions with lower microwave fields. This shows the importance of a uniform microwave field to observe the real decay by charge carrier immobilization, not obscured by other effects.

3.1.4 Surface modification of silicon

In semiconductor technology, the effect of a thin surface layer on the material is very important. This can be for passivation of the surface, generation of a space charge region, or both. SiN deposited on Si obtains a positive charge and consequently induces in weakly doped p-Si an inversion layer. In **Figure 7**, TRMC signals induced by 532 nm light (data points) at the SiN face of a p-Si wafer only at this face covered by SiN are compared to Surface Photo Voltage signals (lines drawn) in a double

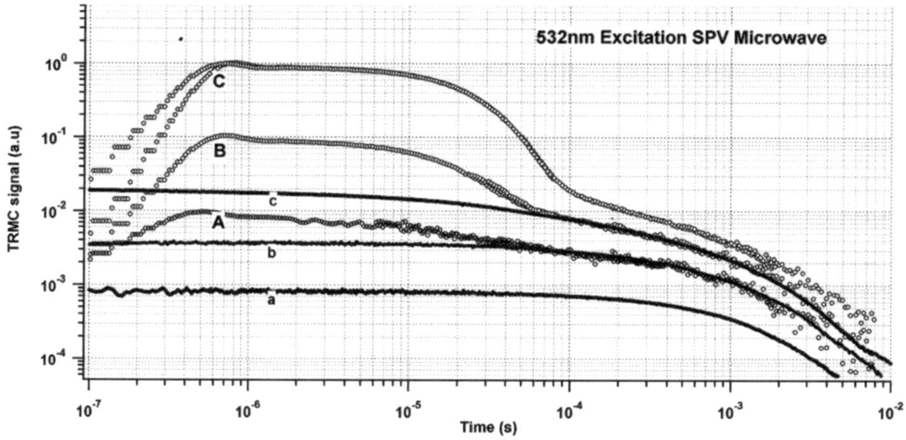

Figure 7.
TRMC transients in a p-Si wafer covered at one face by a SiN film induced by 532 nm light pulses at the covered face at different intensities. The straight line represents surface photovoltage (SPV) signals obtained under the same conditions (courtesy by Phys. Stat. Sol.).

logarithmic representation. SPV signals monitor the change of the voltage at the surface (in this case, over the inversion layer).

The TRMC signals show an initially faster decay followed by a much slower decay. The SPV signals show an equivalent slow decay, but the signal is time-independent before the onset of the slow decay. The initial decay of the TRMC signal represents excess charge carriers outside the space charge region (and so invisible in SPV measurements) and shows decay at the uncovered surface given by Eq. (6).

The slow decay reflects excess charge carriers in the space charge region, generating the photo voltage observed with both methods. Consequently, TRMC measurements can be used to study kinetics in the space charge region.

Interesting is the difference between SiO and SiN covering of silicon. Both obtain a positive charge after deposition on silicon. SiN on silicon has a higher positive charge and so a larger space charge region. However, the volume decay and the decay of charge carriers in the space charge region (i.e., the decay of the photovoltage) is faster in the SiN-covered system.

3.2 Measurements of other materials

In fact, as a contactless method, the TRMC method is most appropriate for materials where the contact properties are largely unknown for all new materials. Also, for materials that cannot easily be contacted as semiconductor powders, the method seems adequate.

3.2.1 Amorphous silicon (a-Si:H)

This material has several applications in electronics. Although the application in p-i-n solar cells is limited because the degradation during extended illumination is handicapped (Staebler–Wronski effect), the use in heterojunctions is still growing strong.

Experiments were performed to verify the applicability of TRMC measurements to amorphous, low-mobility materials. A good agreement with the theory discussed

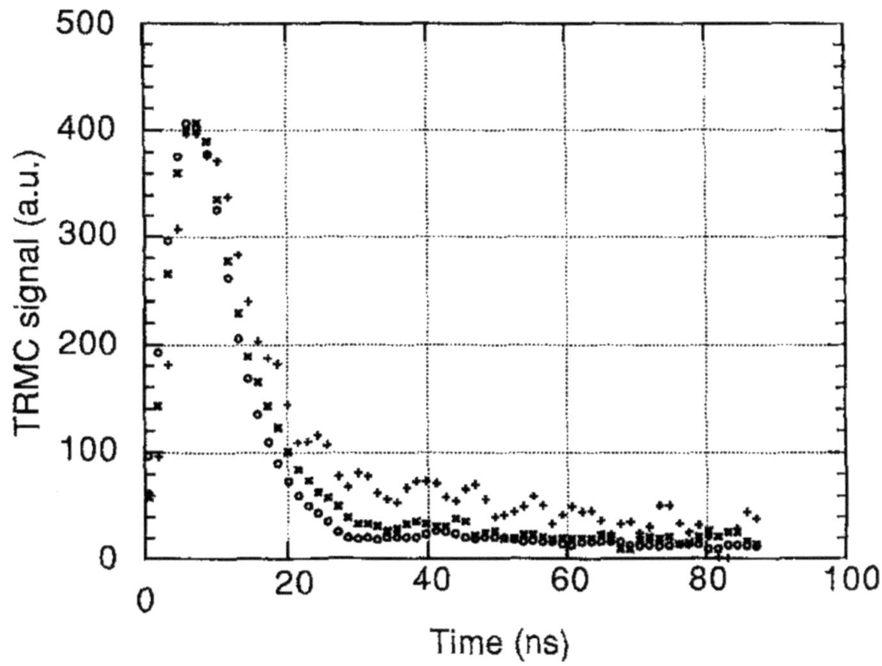

Figure 8.
TRMC transients induced by 532 nm laser pulses in an a-Si:H film at 1.5 mJ/cm² (o), 0.15 mJ/cm² (*) and 0.015 mJ/cm² (+). The maxima of the TRMC amplitudes are normalized at the same value to accentuate the difference in decay behavior (permission by Appl. Phys. A).

above with only mobility as a varying parameter. A mobility of about $0.5 cm^2V^{-1} s^{-1}$ is obtained [8].

The transient photoconductivity in intrinsic and n-doped a-Si:H is due to electrons. The electron mobility is two orders of magnitude larger than the hole mobility, and in intrinsic a-Si:H the Fermi level is nearer to the conduction band (or in other words, it is weakly n-doped). But, even the electron mobility is only about unity, and TRMC measurements must be performed at high laser intensities.

The TRMC signal in a-Si:H after 532 nm excitation follows the pulse shape, giving way to a tail. The importance of the tail relative to the signal maximum at about 10 ns (the TRMC amplitude) increases with decreasing excitation density (**Figure 8**).

The TRMC amplitude increases with the square root of the excitation at high excitation densities and changes in a linear dependence at low excitation densities. The square root dependence is due to a second-order electron-hole recombination already active during the excitation (**Figure 8**).

It appeared that even at the high excitation densities required a reliable impression of the quality of a-Si:H is obtained from the TRMC amplitude [9]. The TRMC amplitude is smaller for lesser quality material due a decrease of the mobility and trapping.

3.2.2 TiO_2 powder

Titania (TiO_2), a white nano-crystalline size powder, has several applications:

DOI: http://dx.doi.org/10.5772/intechopen.1002687

- as a white paint, in dyes, esthetical products, etc.
- as catalyst, in particular for depollution
- in solar cells, covered by a dye to overcome the too-large bandgap for solar energy conversion.

TRMC measurements can elucidate the influence of surface modifications, which is important for these applications.

In **Figure 9** TRMC signals are shown in a double logarithmic representation. The transient b is obtained for an untreated TiO_2 (P25) sample. Treatment of the power with 2-propanol leads to a large decrease of the decay (transient a). On the contrary, treatment with tetranitromethane leads to a very large increase in decay (c). This is explained by increased electron immobilization on the tetranitromethane modified surface and an increased hole trapping at the 2-propanol modified surface.

To explain these effects it must be taken into account that the signal is due to mobile electrons. This explains the faster decay at the tetranitromethane treated surface. The trapping of holes by the 2-propanol treated surface leads to a decrease of the electron-hole recombination and so a slower electron decay and a slower TRMC decay. This experiment shows the sensitivity of TRMC measurements to the influence of surface modification on excess charge carrier kinetics. This makes the technique an important tool for the investigation of photocatalysis on semiconductor powders.

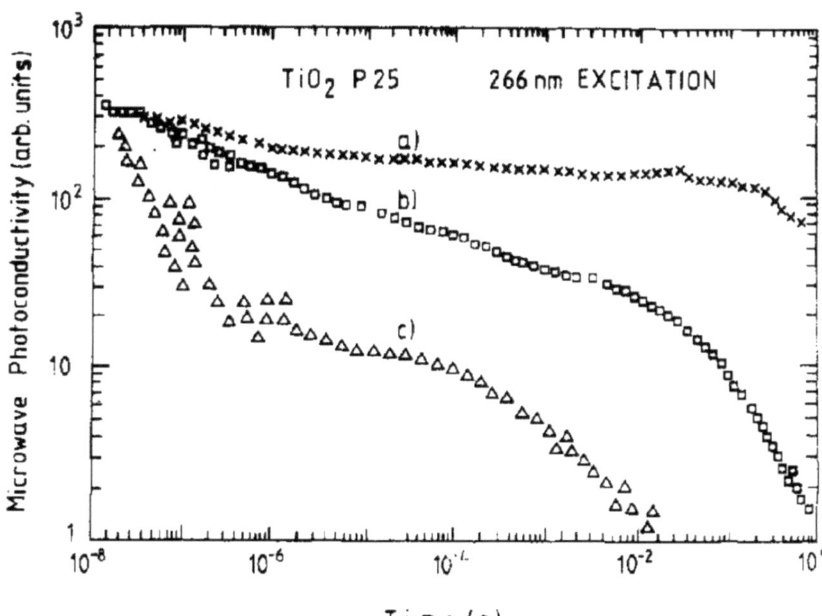

Figure 9.
TRMC signals induced by 266 nm laser pulses in TiO2 (P25) powder, (a) untreated powder, (b) treated with 2-propanol, and (c) treated with tetranitromethane (courtesy by J. Phys. Chem.).

4. Conclusions

In this work it is shown that contactless measurements in the microwave frequency range yield a reliable determination of (photo)conductivity at the hand of measurements on crystalline silicon; a system with known parameters. As example, the influence of an important passivation of silicon by SiN is investigated, and injection of electrons in silicon is observed, so the positive charge of the SiN interface is explained. This leads to an accumulation layer in n-Si and a depletion/inversion layer on p-Si. Further examples concerned a-Si;H and a semiconducting powder, TiO_2.

The technique is particularly appropriate for new semiconductors with unknown contacting properties [10–12] and for the investigation of powders for photocatalysis on semiconductors [13, 14].

Author details

Marinus Kunst* and Reinhard Schwarz
Physics Department and CeFEMA, Instituto Superior Técnico, IST, Lisbon, Portugal

*Address all correspondence to: kunstm@orange.fr

References

[1] Infelta PP, de Haas MP, Warman JM. The study of the transient conductivity of pulse irradiated dielectric liquids on a nanosecond lifetime using microwaves. Radiation Physics and Chemistry. 1977;**10**:353

[2] de Haas MP. Thesis Leyden University. Delft: Delft University Press; 1977

[3] Jackson JD. Classical Electrodynamics. New York: John Wiley; 1962

[4] Blakemore JS. Semiconductor Statistics. Oxford: Pergamon; 1962. pp. 323-326

[5] Kunst M, Sanders A. Transport of excess carriers in silicon wafers. Semiconductor Science and Technology. 1992;7:51

[6] Hezel R, Schorner R. Plasma silicon nitride. A promising dielectric to achieve high-quality MIS/IL solar cells. Journal of Applied Physics. 1981;**52**:3076

[7] Swiatkowski C, Sanders A, Buhre KD, Kunst M. Charge carrier kinetics in semiconductors by microwave conductivity measurements. Journal of Applied Physics. 1995;**78**:1763

[8] Swiatkowski C, Kunst M. Characterization of amorphous silicon films by contactless transient photoconductivity measurements. Applied Physics. 1995;**A61**:623

[9] Neitzert HC. PhD Thesis. Berlin. 1991

[10] Colombi G, Boshuizen B, Chaykina D, Hsu L, Schreuders H, Savenije TJ, et al. Large polaron conduction, photoconductivity and photochromism in CdO_xH_{3-2x}. Advanced Optical Materials. 2023;**11**:202202660

[11] Zhao J, Caselli VM, Bos M, Boshuizen B, Saqvenije TJ. How deep hole traps affect the charge dynamics and collection in bare and bilayers of methylammonium lead bromide. ACS Applied Materials and Interfaces. 2021;**13**:16309

[12] Guse JA, Jones TW, Danos A, McCarney DR. Recombination dynamics in thin film photovoltaic materials via time resolved microwave conductivity. Journal of Visualized Experiments. 2017;**121**:55232

[13] Azevedoo J, Campbell S, He D, Cornul R, Berlucchi M, Sorgues S, et al. Versatile wafer-scale technique for the formation of ultra smooth and thickness-controlled graphene oxide films based on very large flakes. ACS Applied Materials and Interfaces. 2015;7:21270

[14] Herissan A, Nawfal M, Kunst M, Colbeau-Justin C. Time resolved microwave conductivity. In: Bahnemann DW, Patrocinio AOT, editors. Studying Mobile Charge Carriers in TiO2 Photoactive Particles, Springer Handbook of Inorganic Photochemistry. Switzerland: Springer Nature; 2022

Chapter 6

Prospective Applications of Microwave Reflectarray Antennas to the Design and Fabrication of Future Radio Telescopes

Luca Olmi

Abstract

The size, cost, and complexity of radio telescopes and their instrumentation have grown enormously during the last few decades. The vast majority of new technology employed by radio astronomy has been developed for the industrial market, and the technology in use by today's best radio telescopes is 20 to 30 years old. In particular, the antenna technology has changed very slowly, and thus, an important question is whether novel antenna technologies might bring down the cost of collecting area, since efficiency and accuracy compromises become much more costly for large radio telescopes. In this chapter, we discuss the strengths and weaknesses of specific new technologies, and in particular reflectarrays, which have been developed mainly for remote sensing and satellite communications, that might lead to a great leap forward in the design and fabrication of antennas for radio astronomy.

Keywords: radio telescopes, radio astronomy, reflectarrays, transmitarrays, antennas

1. Introduction

Radio astronomy is currently going through a critical transition, where novel electromagnetic (EM) technologies and approaches are replacing the standard radio astronomical methods for radiation detection. These new technologies are mainly based on the use of arrays, the control and modification of the incident wavefront, the use of "metamaterials," and the continuous improvement of the performance of both the telescope and its auxiliary instrumentation. Historically, the metrics used to evaluate the performance of a filled-aperture (or single-dish) radio telescope is based on three main parameters: sensitivity, angular resolution, and spectral resolution. The classical approach to improve the first two parameters consists of using a larger diameter for the primary reflector, leading to much increased engineering complexity and construction costs. As a consequence of this "static" process, where the technology in use by today's best radio telescopes has not changed significantly, the size, cost, and complexity of radio telescopes and their instrumentation have grown enormously during the last few decades.

In particular, while digital technologies applied to radio astronomy have basically followed Moore's law, antenna technology has changed very slowly, whereas the technology of cryogenic amplifiers has followed a somewhat intermediate improvement rate between the previous two extremes. Therefore, in the development of ever larger and better radio telescopes, with expanded collecting area, field of view (FOV), frequency coverage, instantaneous bandwidth, system temperatures, and spatial resolution, the main bottleneck is constituted by the antenna design and fabrication techniques. In fact, the current paradigm is that, to a first approximation, ten times as much money will buy ten times as much collecting area, all else being equal. In this scenario, it is important to explore the possibilities offered by new technologies and design approaches and analyze how they might be able to improve the capabilities of new and existing telescopes in an engineering and cost-effective way.

The vast majority of new technology employed by radio astronomy has been initially developed for the industrial market, and then, it has been adapted for use with radio telescopes and their instrumentation. During this process, a great deal of design and laboratory work is usually required to learn and incorporate these new technologies. For the antennas, the transfer of new designs and fabrication techniques from the ground and satellite communication industry to radio astronomy has proved to be even slower. This is mainly because technological transfer from one field to another always requires a significant (and costly) development phase and also because of a certain degree of risk associated with this adaptation process. However, new antenna technologies are now sufficiently mature to advocate preliminary studies for their potential applications to radio astronomy, at frequencies higher than a few 100 s MHz.

2. Recent telescope development for radio astronomy

In the late 90s and early 2000s, several trends emerged in the development of filled-aperture radio telescopes, specifically in the case of electrically large (D/λ $\gtrsim 5 \times 10^4$ antennas for the microwave/millimeter wave range (the main components of a dual-reflector radio telescope are schematically indicated in **Figure 1**). These trends concerned both the mechanical and the optical design of the telescope, which also had to accommodate the newly developed heterodyne and bolometric focal plane arrays (e.g., the review in [1]). In particular, the mechanical design was generally focused on achieving the best possible homology. On the other hand, depending on the scientific programs, the optical designs had different prominent features, such as off-axis (e.g., Green Bank Telescope, GBT), wide-field (e.g., South Pole Telescope, SPT), and shaped optical configurations (e.g., Sardinia Radio Telescope, SRT). If we extend our review into more recent years and also to the submillimeter wave range we can note that three main classes of radio telescopes have been finalized or are currently in the design or construction phase (see also **Table 1**): (i) large ($\gtrsim$50 m) single-dish instruments (e.g., LMT, AtLAST); (ii) extended synthesis arrays (e.g., ALMA, SKA, ngVLA); and (iii) smaller instruments designed for specific projects or with specific enhanced performance (e.g., CCAT/FYST).

Despite the great variety in the specific design, overall performance, and scientific applications, all of the telescope projects described above are still based on standard technologies for designing the collecting area of the instruments (in particular at frequencies higher than a few 100 s MHz), with possibly some innovations only in the

DOI: http://dx.doi.org/10.5772/intechopen.1002067

Figure 1.
Example of a classical dual-reflector radio telescope working in the microwave region. Radio waves reflect off the primary and secondary surfaces and focus at the position of the feed horn (or array of feed horns), where the transition from propagation in free space to guided propagation takes place. Receivers and other auxiliary devices then amplify, detect, and analyze the radio signals.

Telescope	Type	Present status	Optical configuration	Diameter of primary reflector (m)	Frequency range (GHz)
SRT[a]	Single-dish	Upgrading	Shaped	64	≈0.3 – 115
			Gregorian		
LMT[b]	Single-dish	Operational	Cassegrain	50	≈75 – 350
JCMT[c]	Single-dish	Operational	Cassegrain	15	≈85 – 350
FYST[d]	Single-dish	Under	Crossed	6	≈210 – 850
		construction	Dragone		
SKA-Mid[e]	Interferometer	Under	Offset	15	≈0.3 – 15
		construction	Gregorian		
ngVLA[f]	Interferometer	Design	Offset	18	≈1 – 115
		phase	Gregorian		

[a]*Sardinia Radio Telescope (http://www.srt.inaf.it/)*[b]*Large Millimeter Telescope (http://lmtgtm.org/)*[c]*James Clerk Maxwell Telescope (https://www.eaobservatory.org/jcmt/)*[d]*Fred Young Submillimeter Telescope (https://www.ccatobservatory.org/)*
[e]*Square Kilometer Array Mid-frequency telescopes (https://www.skao.int/en/explore/telescopes/ska-mid)*[f]*Next Generation Very Large Array (https://ngvla.nrao.edu/)*

Table 1.
Technical characteristics of some representative radio telescopes currently in operation or in the design/construction phase. For arrays of antennas, the properties of the individual antenna are shown.

fabrication of the surfaces themselves, such as hydroforming. Most of the design efforts were instead concentrated in the development of active surfaces and metrology systems, improved optical designs, and the use of novel instruments at the focal plane of the telescope, such as focal plane arrays and phased array feeds. Thus, most of the technological innovations have gone in the front-end and back-end development and have resulted in improved frequency coverage, FOV, instantaneous bandwidth, and system temperature.

We may note that while optical telescopes have indeed been characterized by the use of a novel technology to design and build the primary reflector, specifically, the International Liquid Mirror Telescope (ILMT[1]), to our knowledge the only novel and funded concept for a radio to submm telescope design has been the Large Balloon Reflector (LBR[2]), where the main reflector of the telescope is a metallized hemisphere of a smaller spherical balloon, 5 to 20 m in diameter, inflated inside the much larger (100 m) carrier stratospheric balloon. Although the LBR represents a game-changing approach to realizing large aperture telescopes at THz frequencies, this project was specifically designed as a balloon-borne telescope and cannot be easily adapted to build ground-based telescopes.

Even national, or international consortia borne to foster the development of radio astronomy, such as RadioNet[3] in the past, and more recently Radioblocks[4] that has been specifically proposed and funded to develop next-generation technologies for radio astronomy infrastructures, will mainly focus on the development of new front-ends and back-ends, but not on antenna technology. Given that the cost of large instruments for radio astronomy is largely dominated by the antennas, both during the design/construction phase and later in terms of maintenance costs [2], this leaves the most impacting component in the design and fabrication of a radio telescope without an adequately sustained technology R&D program. Since the current trend in the development of large-scale international radio telescope projects is to explicitly include cost in the design process, it becomes extremely important to explore the parameters space based on the use of new technologies for the design and fabrication of the antennas.

3. Recent and emerging technologies

During the last 10–20 years, great efforts have been carried out in the radar and communications industry to control and manipulate the phase (and, in some cases, also the amplitude) of the incident and reflected, or transmitted, waves. The use of these phase controlling devices is likely to continue to expand in the near future, also for photonics applications. At the basis of this technological development lies the concept of the *metamaterials*, which are three-dimensional (3D), often periodic, artificial materials composed of metals and/or dielectrics, with the ability to modify waves beyond the capabilities of naturally occurring or homogeneous materials (e.g., the review in [3]). Their exceptional ability to manipulate waves is due to their interaction with electric and/or magnetic fields, which can be controlled by the

---

1 http://www.ilmt.ulg.ac.be/ilmt

2 https://www.nasa.gov/content/10-meter-sub-orbital-large-balloon-reflector-lbr-1/

3 https://www.radionet-org.eu/radionet/

4 https://www.jive.eu/radioblocks-eu-next-generation-technologies-radioastronomy

DOI: http://dx.doi.org/10.5772/intechopen.1002067

geometry of the individual artificial scatterers, and leads to a wide range of applications such as antenna performance enhancement, perfect absorbers, superlenses, in a variety of wavebands from the microwave to the optical range.

Metasurfaces are two-dimensional (2D) or planar versions of metamaterials with subwavelength thickness, and consist of thin artificial layers with periodic arrangements of small inclusions in a dielectric host medium. They are characterized by light weight and ease of manufacturing, and by engineering the metasurface with different properties to achieve specific functions, numerous applications are exploited in electromagnetics ranging from microwave to terahertz (THz) frequencies and even to optical frequencies, for example, sensors, lenses, absorbers, and reflectarrays. Different functions in these applications are achieved by different unit cell geometries with required reflection or transmission properties. Depending on the specific design and application, and the size of the unit cells, metasurfaces are also sometimes known as phase-shifting surfaces (PSS, [4]), frequency-selective surfaces (FSS), intelligent reflecting surfaces (IRS), or Huyghens' surfaces [5]. All applications of metasurfaces mentioned above have intrinsic fundamental limitations including narrow bandwidth, linear response, fixed functionality, etc., that will be discussed later. However, recent studies on active and nonlinear metasurfaces have enabled the design, fabrication, and test of metasurfaces with nonlinearity, power dependency, tenability, and switching abilities. This has been achieved by applying active electronics like transistors, diodes, and varactors to traditional passive structures.

Metasurface applications of interest to radio astronomy are mainly those related to wavefront engineering such as beam shaping, including focusing, defocusing, reflection, and refraction [5]. This class of metasurfaces is most commonly known as reflectarrays and transmitarrays (see **Figure 2**). A reflectarray is a flat, planar, compact structure, typically illuminated by a feed, consisting of an array of isolated radiating elements that are predesigned with a particular phase delay to re-radiate and scatter the incident field to form a desired wavefront. The individual resonant unitary elements are called unit cells and are predesigned with a particular phase delay to scatter the incident field to form the desired wavefront in front of the aperture. Thus, although they are physically flat, reflectarrays behave electrically as a curved reflector. Reflectarrays are manufactured by photo-etching and bonding processes, which are a well-established technology in multi-layer printed circuit boards.

A transmitarray (or array lens antenna) is also composed of an array of multiple discrete elements used as a transmitting device. The name is inspired by the reflectarray, but the device is used as a transmitting surface rather than a reflective one. When made out of printed metallic elements, both the reflectarray and the transmitarray are physically flat. Just as a reflectarray can emulate a conventional reflector, a transmitarray can emulate a conventional lens. However, a single layer is generally not enough to realize the needed range of phase shift, and thus, many layers need to be used in cascade. Although the transmitarray does not allow for achieving an optimum thickness reduction, it can be significantly thinner than a standard dielectric lens. Since in reflectarrays the feeding source is on the same side of the radiated field, for similar applications transmitarrays are generally preferred to reflectarrays in all those cases where feed blockage may seriously affect the final performance of the device.

Therefore, reflectarrays and transmitarrays combine the advantages of reflector antennas (or lenses) and phased arrays, and they can replace the traditional solid optical elements, such as parabolic/hyperbolic reflectors and lenses, with easy to handle, low profile, and lightweight planar structures. In addition, since they employ

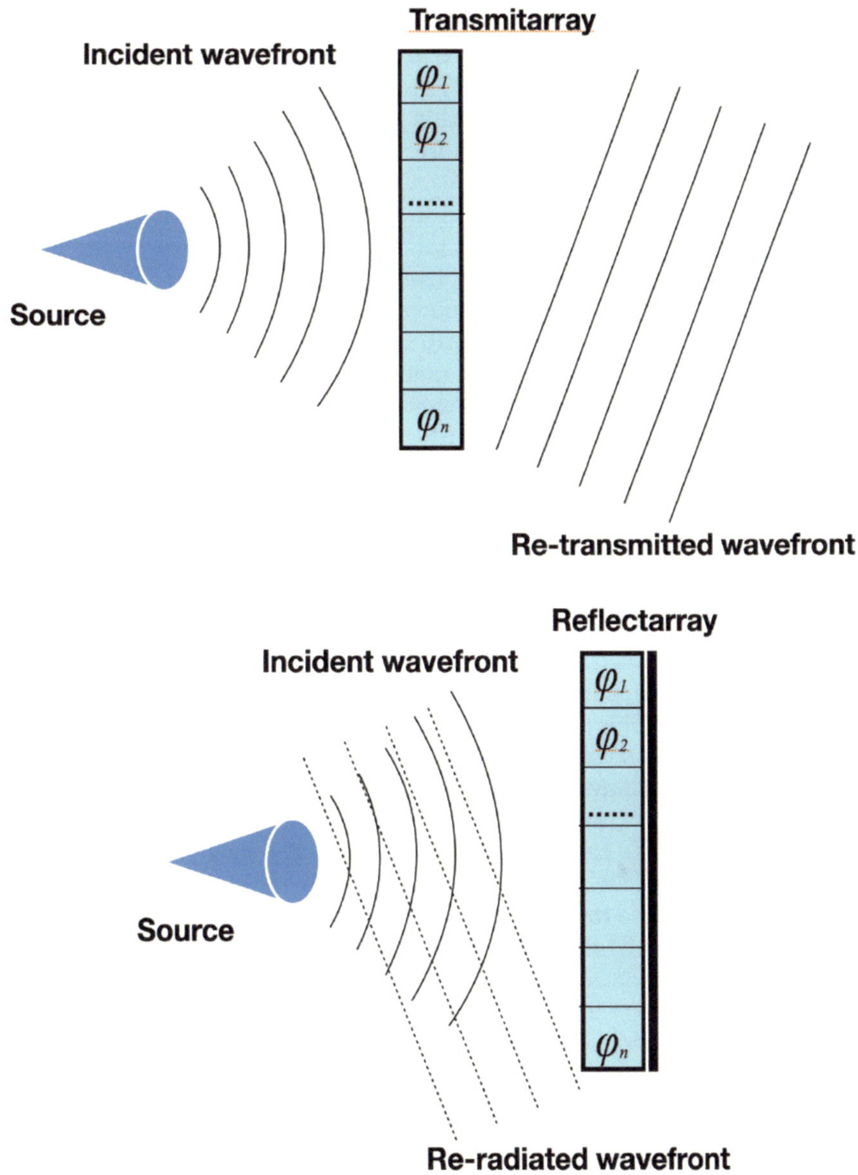

Figure 2.
Generic model of a transmitarray (top panel) and a reflectarray antenna (bottom panel). $\{\phi_1, \phi_2, \ldots, \phi_n\}$ *indicate the phase delays applied to the incident wavefront by each unit cell to scatter the incident field and form the desired wavefront.*

spatial feeding instead of using traditional feed networks, they enable a significant reduction in system complexity and cost, particularly when there is a need to use a large number of unit cells. There are, however, a number of aspects that are specific to metasurfaces in general, such as the need to use low-loss, but low-cost, substrate materials, and the requirement to work and survive a wide range of weather conditions, which may require appropriate environmental protection.

DOI: http://dx.doi.org/10.5772/intechopen.1002067

The initial design of reflectarrays and transmitarrays was characterized by a fixed-beam (also known as a passive type). In a fixed-beam array, the beam forming is implemented by properly adjusting the dimensions at each unit cell and cannot be modified. Later, arrays were also developed where the beam could be dynamically reconfigured or scanned by introducing controllable mechanisms at the element level, in order to change the phase-shift and to reconfigure the beam. These are also known as reconfigurable (or active) reflectarrays and transmitarrays, and they are of particular interest for radio astronomical applications since pointing and tracking can in principle be performed electronically, with no or few moving mechanical parts. On the other hand, reconfigurability is achieved at the cost of increased complexity of analysis, synthesis, phase-shifter design and technology, calibration, control, and testing [6, 7].

4. Potential applications to radio telescopes

As mentioned earlier, the main issue that we would like to address in this work is weather the novel technologies described in the previous section are sufficiently mature for applications to radio telescopes. However, we will see that there is not a unique answer to this question, since this depends on a wide range of constraints including (but not limited to) the frequency range, size of the antenna(s), and the specific scientific application. Therefore, our discussion will be limited to some more general aspects of these novel technologies that are likely to affect their application to radio telescopes in the near future, such as:

- maximum size of the collecting area;
- multi-band applications and bandwidth limitations;
- beam-steering capabilities and limitations;
- applications as single- or dual-reflector antennas.

Our discussion will also be limited to the microwave and millimeter wave range. In fact, while reflectarray techniques developed for the microwave range can be in general extended also to higher frequencies, several factors emerge that complicate the antenna design, which are often related to material properties and also to cost.

4.1 Collecting area

While certain scientific applications may require telescope designs optimized for specific parameters (e.g., FOV, system temperature), large collecting areas are still desirable for single-dish telescopes that have a wide range of scientific goals. Even for telescope arrays, the size of the individual antenna is a fundamental parameter in the critical trade-off between performance and cost. Therefore, it is important to address the question of how large reflectarrays and transmitarrays can be fabricated. Given that the unit cell has typically a size $\approx 0.5\lambda$, the number of elements grows quickly with the electrical diameter, D/λ, of the antenna. One of the largest planar reflectarray antennas ever fabricated is a 3-m Ka-band inflatable reflectarray developed at NASA for space applications, which is composed of approximately 2×10^5 elements. A 3-m dual-band Cassegrain reflectarray (X/Ka-band) with a reflectarray main reflector was demonstrated in Ref. [8]. This reflectarray was composed of 19,000 and 275,000 elements at X-band (8.4 GHz) and Ka-band (32 GHz), respectively.

While even larger reflectarray could certainly be built, it is currently impractical to fabricate antennas for the microwave range with a collecting area of 10s or 100 s of square meters with a single reflectarray (and even less a transmitarray, see Section 4.6). Besides to complications with the technologies used for the fabrication of the reflectarrays, an important limitation to the design of large reflectarrays is the need to compensate for the differential spatial phase delay, which increases with the size of the aperture and also contributes to limiting the bandwidth (see Section 4.3). Another factor that may severely affect the maximum size of a single reflectarray antenna is whether it is designed to have beam-steering capabilities.

In general, a reflectarray antenna mimics the optical behavior of a parabolic reflector, with the advantage of having a flat surface instead of a curved reflecting surface. However, in order to build large reflectarrays it may be necessary to use radiating elements distributed on a curved surface (also known as conformal reflectarrays), or also using multi-facet, or piecewise, configurations that approximate the shape of a parabolic surface using flat reflectarray panels. Both design options introduce a geometric compensation to the feed-element paths delays. In these cases, however, one of the greatest advantages of standard reflectarrays, their flat structure and ease of manufacturing, must be abandoned, at least partially. The added complexity of the design, fabrication process, and mechanical structure must be weighed against the increased collecting area and performance of the reflectarray. Therefore, we may safely assume that the application of reflectarrays and transmitarrays to radio telescopes will be limited, in the near future, to relatively small apertures (with an area $\lesssim 10\ m^2$ in the microwave range), that will be operated as individual antennas of an array, as part of a larger multi-facet antenna or also as secondary reflector in a dual-reflector configuration, as discussed in Section 4.5.

4.2 Beam-scanning capabilities of reflectarray antennas

Pointing and tracking with radio telescopes in the microwave range is achieved by a fully-steerable mechanical structure (or by the movement of the subreflector alone for small angular ranges, up to a few beams), the use of phased array feeds (PAF), or also a combination of both. In fact, this might be the case for the SKA1-Mid antennas, which have been designed to incorporate PAF receivers in the future. The choice between these options is usually driven by factors relating to the mapping speed, FOV, and cost. The electronically scanned PAFs rely on phase shifter technology and although several such receivers have been developed during the last $\sim$ 10 years, their main drawbacks still remain the hardware complexity, their broadband sensitivity, and high cost, which may significantly affect their prospect to be built in large numbers for arrays of antennas.

As discussed above, reflectarrays combine key features of standard reflectors and phased array elements to generate a collimated beam as required in high gain antennas, including radio telescopes operating at frequencies higher than a few 100 s MHz. In addition to the various advantages of reflectarrays over standard reflectors for fixed-beam applications, the reflectarray antenna can also establish a new paradigm for beam-scanning applications. The beam of a reflectarray antenna can be scanned by

DOI: http://dx.doi.org/10.5772/intechopen.1002067

means of the reflector nature or the array nature of the antenna. In the first approach (also called feed-tuning technique), passive reflectarrays are used and it is the feed phase center, which is mechanically moved along a specified path, thus achieving a corresponding angular shift of the beam on the sky, as with a standard reflector. In the second approach (also known as aperture phase-tuning technique), the individual elements on the reflectarray aperture are equipped with a phase-tuning mechanism in order to change the phase-shift and to reconfigure the beam, thus utilizing the array nature of the reflectarray antenna [7, 9]. Clearly, it is also possible to utilize both approaches in a single design to improve scan performance or reduce system cost. Reflectarrays can be fabricated as the subreflector, main reflector, or both in dual-reflector configurations. The choice of single- or dual-reflector configuration is fundamental in determining the beam-scanning performance of the antenna, and it is even more important for the feed-tuning technique since the reflectarray has no phase-tuning capability in this configuration.

Many examples of beam-scanning passive reflectarrays utilizing the feed-tuning technique can be found in the literature (see the review in [9]), with both parabolic and non-parabolic phase distributions over the array, but most of them can achieve a scanning range of a few or several beams. Thus, while this technique has the advantage of the ease of fabrication for the passive reflectarrays, it may not be the optimum choice for radio astronomical applications, unless other beam-steering techniques are also used to increase the FOV. For applications where wide-angle scan coverage is required, including radio telescopes, the aperture phase-tuning approach that utilizes the array nature of the reflectarray antenna is the more promising choice. However, while in passive reflectarray configurations the reflection phase for each element is a fixed quantity, for reconfigurable reflectarrays a tunable phase-shifting mechanism must be incorporated into the individual elements, as already mentioned. Different approaches are available for tuning the phase of reflectarray elements, and each enabling technology has different advantages and limitations, but they all share an increased level of complexity and cost.

The aperture phase-tuning technology is still in its development stage, and many advanced design concepts are only recently being explored. However, its continuous development and diversified applications are likely to lead in the near future to systems that will be able to replace the standard techniques for beam-steering in radio astronomy. Scan performance could also be improved by combining different scanning methodologies, such as multi-facet configurations [10], spherical-phase reflectarray antennas (SPRA, [11]), and the use of dual-reflector configurations with a sub-reflectarray and a spherical primary [12].

The scan capabilities of active reflectarrays are also closely related to their bandwidth performance (see Section 4.3). In fact, the development of new strategies able to improve the bandwidth of active unit cells is an ongoing challenge in the reflectarray antenna design, since most existing active configurations are characterized by only a few percent bandwidths. A review of the bandwidth performance of reconfigurable reflectarrays can be found in Ref. [13].

4.3 Reflectarray bandwidth limitations

Modern radio telescopes are required to operate in a wide range of frequencies, often covering more than an order of magnitude. In addition, state-of-the-art radio astronomical receivers can cover a large instantaneous fractional bandwidth, typically ~ 30 % or higher. As described in the previous sections, reflectarrays mimic the standard reflector antennas, but there is an inherent fundamental difference between the two regarding their bandwidths. A reflector antenna is essentially frequency-independent; that is, it has an infinite bandwidth, which is instead limited by the feed and the receiver. On the other hand, reflectarrays suffer from bandwidth limitations due to the intrinsic narrow-band of microstrip radiators and also the differential spatial phase delays from the feed to the reflecting elements. The second effect is dominant for reflectarrays with large aperture sizes D and small focal distances f from the feed (see **Figure 3**), resulting in $f/D \lesssim 1$. On the other hand, in the case of reflectarrays with $f/D \gtrsim 1$, the dominant factor limiting the reflectarray bandwidth is the unit cell operating band [13].

Both factors introduce phase errors as a function of frequency, which limit the bandwidth of reflectarray antennas. Many solutions have been presented in the literature to enhance the operating band of passive reflectarrays, and improved bandwidths up to ≈20 % have been demonstrated in several designs. However, improving the bandwidth of active unit cells is still an ongoing challenge in the reflectarray antenna design. In fact, most existing active configurations are characterized by few percent bandwidths [13–15].

Regardless of the choice of the unit cells used for the design, the overall bandwidth will still be limited by the flat nature of the reflectarray aperture. This limiting factor, due to the differential spatial phase delay, can be explained by referring to **Figure 3**, where for simplicity we consider the case of a broadside pencil-beam reflectarray, consisting of an axial-fed reflectarray designed to have a focal distance f. The reflectarray elements are supposed to be in the far-field region of the feed, and thus, we can assume that a spherical wave is incident on the reflectarray aperture. For this simple case, the reflectarray must be designed to convert the incident spherical wavefront into a reflected plane wave propagating on-axis. This can be achieved by considering the quantity $\Delta R(x, y)$ representing the path length difference between the

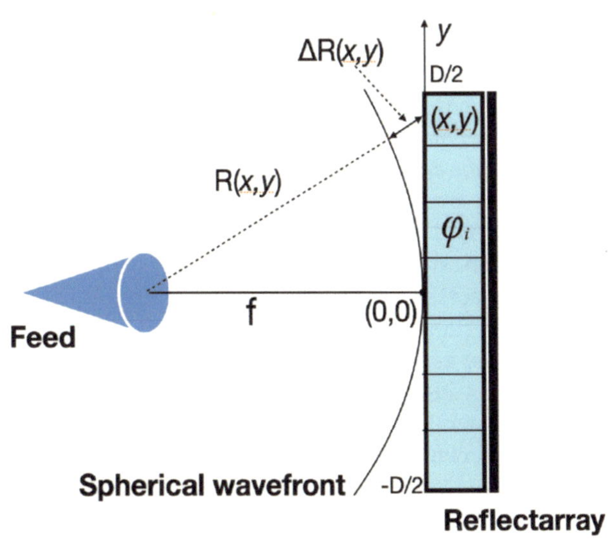

Figure 3.
Reflectarray geometry and the differential spatial phase delay. $\Delta R(x, y)$ represents the path length difference between the nominal focal length, f, and the distance from the feed to each element of the reflectarray at position (x, y).

required focal length and the distance from the feed to each element of the reflectarray at position (x, y). This path length difference generates a frequency-dependent phase difference, or differential spatial phase delay, given by:

$$\phi(x, y, \nu_\circ) = \frac{2\pi}{c} \nu_\circ \Delta R(x, y) \qquad (1)$$

For the reflectarray design, the phase shift must be adjusted in each element to compensate for the phases given by $\phi(x, y, \nu)$. In receiving mode, a plane wave incident on the reflectarray along the boresight direction will be focused on the feed, thus mimicking the response of a paraboloidal antenna. The reflectarray elements are usually selected to satisfy the required phase at the design frequency $\nu_\circ$ As the frequency changes, the phase delays in the array will change, and the phase response of the unit cell will also change. These two phase variations do not generally match with each other, thus introducing phase errors that will result in pattern deterioration and as a consequence bandwidth limitations of the system.

For beam-scanning applications, the differential spatial phase delay combines the spatial delay from the feed to different elements on the surface (thus increasing with the size of the reflectarray), and the variation of this differential delay due to the incident radiation coming from different angles (thus increasing with the required angular scanning range). The effect on the bandwidth of the compensation mechanism for different phase delays is thus dominant in the case of large reflectarrays and/or wide beam scanning ranges. Therefore, the development of techniques able to exactly compensate the differential phase delay in a large frequency range would be required for radio astronomical applications [16]. An interesting technique that can in principle achieve beam-steering and increase the instantaneous bandwidth is the use of reflectarrays with orbital angular momentum (OAM)-based structures [17].

4.4 Multi-band operations

Besides wide instantaneous bandwidths, radio telescopes must generally provide the coverage of two or more widely separated frequency bands. Multi-band operation of reflectarray antennas can be implemented by the assembly of several sub-arrays, each operating in a different waveband. These configurations can be generally achieved using two or more stacked layers, with the drawback of an increased cost and complexity [15]. However, single-layer multi-band reflectarrays are more easily manufactured, have a much lower cost, and do not face problems associated with the precise layer alignment. The specific technique to achieve multi-band performance depends on the number of frequency bands and their frequency ratios. Several multi-band reflectarray prototypes, without beam scanning capabilities, have been realized, combining up to six different frequency bands, mostly in the range from C-band to Ka-band [18]. Current research in this specific area is focusing on designing unit cells that minimize the mutual coupling effects between the elements at all frequency bands.

In addition to reducing these coupling effects, radio astronomical applications would ideally require reflectarray unit cells that are designed to simultaneously implement the beam-scanning function (i.e., reconfigurability) with multi-band capabilities. A first proof-of-concept of a multi-band beam-steering reflectarray is reported in Ref. [19], which describes a dual-band cell in K and Ka bands. Another example of a reflectarray cell combining reflection phase tunability (i.e., beam-steering) with frequency

Device type	Mode of operation	Frequencies (GHz)	Fractional bandwidth (%)	Number of elements	References
Reflectarray	Fixed beam	$\sim 5 - 300$	$\lesssim 20$	$\sim 10s - 10^5$	[7, 15]
Reflectarray	Reconfigurable[a]	$\sim 5 - 60$	$\lesssim 7$	$\sim 10s - 10^4$	[7, 13, 15]
Reflectarray	Multi-band	$\sim 6 - 180$	$\lesssim 6$	$\sim 10s - 10^4$	[15, 18]
Transmitarray	Passive/Active	$\sim 5 - 60$	$\lesssim 30$	$\sim 10s - 10^3$	[22, 23]

[a]*The beam-scanning range is currently limited to $\lesssim 10$ beamwidths with either the feed-tuning or aperture phase-tuning techniques.*

Table 2.
Key features of microwave reflectarrays and transmitarrays as reported in the literature.

reconfigurability is described in Ref. [20]. In this specific case, frequency reconfiguration did not increase the instantaneous bandwidth of the reflectarray, and thus, such a system is useful for narrow to moderate bandwidth applications whose operating band changes over a much wider frequency range.

Ideally, for radio astronomical applications reflectarray cells should be properly designed to have acceptable instantaneous bandwidths and simultaneously implement the beam-scanning function with multi-band operation. The proposed concepts found in the literature are specifically designed for applications in radar and communications systems, while no design exists yet for applications with radio telescopes. The trade-offs and optimization required may also depend on the specific telescope configuration and scientific goals. The parameters space is so wide and complex that the specific reflectarray design cannot be conducted *via* analytical methods; thus, evolutionary optimization algorithms, or other similar optimization techniques, are likely to be required (see, e.g., [21]). A summary of the main features of reflectarrays (and transmitarrays) is shown in **Table 2**.

4.5 Optical configurations for radio telescopes employing reflectarrays

4.5.1 Single- and dual-reflector configurations

In the literature, it is possible to find several papers showing that the main, the secondary, or both reflectors in a dual-reflector configuration can be effectively replaced by reflectarrays. Various dual-reflector antennas with a reflectarray as the subreflector (or sub-reflectarray) have been proposed in the literature [24–26]. In this technique, a reflectarray acts as a flat subreflector and by properly adjusting the phase shifts of the unit cells, they can mimic an ellipsoidal- or hyperboloidal-type subreflector for a dual reflector system, or also perform actively phase compensation. If also the main reflector is replaced by a flat reflectarray, a dual-reflectarray antenna is obtained [27–29]. This configuration provides phase control on both surfaces, which can be used for different purposes such as amplitude and phase synthesis.

In the Cassegrain dual-reflector antenna employing a flat reflectarray subreflector, as analyzed for example in Ref. [24], the antenna beam can be scanned by introducing an appropriate progressive phase distribution across the reflectarray surface. This configuration is attractive for radio telescopes, because it combines the high gain and broad bandwidth properties of the parabolic main reflector with the simplicity of manufacturing a relatively small electronically reconfigurable sub-reflectarray.

The achievable angular range with this configuration can be increased by using a more elaborate phase-synthesis technique to obtain the phase distribution required on the sub-reflectarray for each beam pointing [30] (see also the review in Ref. [27]). However, the ultimate scanning range that can be achieved by electronically reconfiguring the sub-reflectarray may not be sufficient for most radio astronomical applications and thus, the antenna would still need a fully steerable mechanical support. The use of a dual-reflectarray antenna, with planar reflectarrays of which only one or both can be reconfigurable, can add additional features to the antenna system. The only dual-reflectarray antennas that can be found in the literature were designed for satellite communications systems, and thus, a preliminary study for radio astronomical applications had not been done until very recently (see Section 4.5.2) [31].

All of these systems use the conventional reflectarray antenna design, which is based on the phase compensation of a comparable parabolic reflector antenna (thus, they are also known as parabolic-phase planar reflectarrays), for the main reflector, or of a hyperbolic (ellipsoidal) surface for the subreflector in an equivalent dual-reflector Cassegrain (Gregorian) configuration. These antennas are normally designed for a fully illuminated aperture, leading to high aperture efficiency, but also a limited scan coverage, as we previously mentioned. On the other hand, wide-angle scan coverage with reflector antennas can be achieved by using non-parabolic reflector configurations such as SPRAs, where different portions of the reflector surface are illuminated to collimate the beam in different directions [11, 32]. In this configuration, in order to minimize the adverse effects of the spherical aberration only a small portion of the aperture is illuminated for each pointing position, and thus, the total aperture size will depend on the required scan range. This configuration is attractive for radio astronomical applications, also compared to the use of a fixed real spherical surface with a sub-reflectarray (discussed below), because it only needs planar reflectarrays. However, a detailed comparison should take into consideration also the required frequency range and corresponding bandwidths.

In Section 4.1, we mentioned that multi-facet, or piecewise, configurations introduce a geometric compensation to the feed-elements path delays that allow larger reflectarrays to be designed and in turn can also increase their bandwidth performance. An interesting design of this type is that discussed in Ref. [14] where the authors show that by using a reflectarray with a central rectangular panel and corner trapezoidal panels arranged in a concave shape, it is possible to remove the phase error caused by incident wave variation. The other benefit of their design is a significant increase of the system bandwidth.

4.5.2 Applications to radio telescopes

In Ref. [31], the author analyzed some of the optical configurations incorporating the reflectarray technology described in the previous sections. The purpose of that work was to determine which optical configuration is best suited to exploit the beam-scanning capabilities of these devices, and also discuss the possibility of eliminating the need for any mechanical pointing/tracking system. In this preliminary work, the actual reflectarray networks were not simulated, and only EM simulations of standard antennas were used to evaluate the phase distributions that would be required by ideal reflectarrays.

Both single- and dual-reflector configurations were simulated by the author. Specifically, it was shown that a single-reflector telescope, with a large focal ratio (>1 and especially $>>1$) and composed by many individual reflectarrays arranged in a

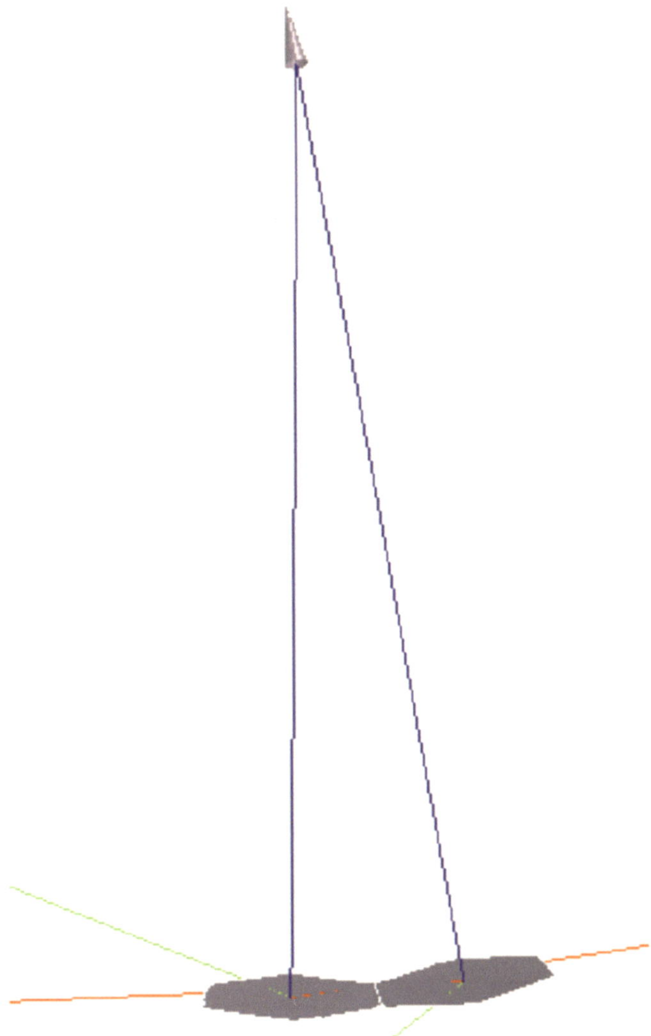

Figure 4.
Example of a reflectarray piecewise configuration using hexagonal panels, each one reproducing the phase distribution of a parabolic reflector with $f/D = 5$. Also shown is the feed at the focus of the antennas, oriented to illuminate the tilted reflectarray panel. The hexagonal profile has been selected because it allows an easy expansion of the total collecting area.

piecewise configuration (see **Figure 4**), similar to that described in Ref. [14], would indeed be able to extend the angular scanning range of the telescope with minimum mechanical movement of the feed support system. If the piecewise configuration is realized in one axis only (e.g., to cover the required elevation range), then the only mechanical movement required would be in azimuth, which is less demanding in terms of engineering requirements. Even if some partial mechanical movement may still be necessary, these solutions would still benefit from the low-profile, low-mass, and low-cost features of the reflectarray technology.

A few simulations with a dual-reflectarray configuration were also carried out in Ref. [31], and the results suggested that the telescope would indeed be able to point and track a target within a small angular field ($\lesssim$10 beams). On the other hand, a large scanning range for astronomical applications would still require the mechanical movement of the antenna. The advantages of the dual-reflectarray configuration are the compact mechanical structure, as in the classical Cassegrain (or Gregorian) optical configuration, and also the fact that the sub-reflectarray would be composed by a much smaller number of individual phase-changing elements, compared to the larger area of the primary surface. However, the work discussed in Ref. [31] is far from being exhaustive since other promising configurations, such as a dual-reflectarray configuration with a multi-facet primary reflectarray, or alternatively using a non-parabolic illumination, were not considered.

Based on the few models that were simulated in Ref. [31] the author concluded that the most promising solution, providing the largest angular range with minimum mechanical complexity, would be a spherical primary reflector with either a single moving sub-reflectarray/feed support or a system composed by multiple sub-reflectarrays/feeds that would only require an azimuthal movement. In the latter configuration, only the feed needs to be moved, contrary to the antenna azimuthal movement previously mentioned. This solution is therefore well suited for antennas with very large collecting areas, up to several 1000s and 10,000 s square meters (see also [33]). As it was discussed above, SPRAs are another competitive design to achieve a large angular scanning range.

4.6 Transmitarray antennas

As previously mentioned, transmitarrays can modify the original radiation pattern of a directional antenna source, for example, a horn antenna, after transmission through the array (see **Figure 2**). A transmitarray thus acts like a lens, allowing to pass-through the incident wave while modifying its direction of propagation. The direction to which the incident wave is being re-radiated depends on the design of the structure. These structures are composed of several resonant unitary elements with a spatial periodicity forming a planar array, in a way very similar to reflectarrays. The fundamental difference between a transmitarray and reflectarray is that in the former both magnitude and phase control of the element are required. In fact, while in a reflectarray all power is re-radiated, independent of the frequency or cell design (and thus only phase control of the unit cell is required), and in transmitarrays the magnitude of the transmission coefficient needs to be close to 1, to ensure a high efficiency over the entire operational bandwidth(s). In fact, if the unit cells are not adapted to the frequency of operation the incident, EM wave will be totally reflected back, resulting in no transmission through the structure. Therefore, a transmitarray is desirable to be as transparent as possible, introducing very low loss so the EM field of the propagating wave is not severely attenuated [22].

The challenges for designing electronically reconfigurable beam-forming transmitarrays are similar to those described earlier for reflectarrays. However, the designs currently found in the literature may be appropriate only for small transmitarrays (from a few 10s up to several 100 s elements) or applications where only a small scanning range is required. Thus, they look less promising for radio astronomical applications than reflectarrays, except maybe for those specific applications where feed blockage is not desirable, such as possibly in multi-feed, multi-beam applications [34]. A review of the characteristics of various beam-steering high gain reconfigurable transmitarrays and their bandwidths can be found in Ref. [23]. An interesting example is a 200-mm-diameter refractive telescope composed of metamaterial Gradient Index (GrIn) lenses based on photolithographic meshes, which has been developed for space applications at sub-THz frequencies [35]. This prototype has a wide FOV ($\pm 10^{\circ}$) and can operate at frequencies between 55 and 183 GHz. The drawback of this specific device was the low optical efficiency and the inadequacy of the design to build a much larger diameter telescope composed solely of metamaterials. However, this result is of scientific interest for its potential applications as an individual optical component of a telescope with a much larger collecting area.

5. Conclusions

The size, cost, and complexity of radio telescopes and their auxiliary instrumentation have grown enormously in the recent decades. Most of the newest design and fabrication methods applied in radio astronomy were originally developed for industry, and some technologies used today in the best radio telescopes date back to 20–30 years ago. In particular, the methods and technologies used in the design and manufacture of the reflecting surfaces of radio telescopes have remained almost unchanged during all this time. A fundamental question for the future of radio astronomy is therefore whether the antenna technologies currently under development could reduce the cost of the collecting area and almost completely eliminate complex mechanical support and moving structures. In this work, we have analyzed some emerging technologies, developed mainly for remote sensing and satellite communications, which could lead to a paradigm shift in the design and manufacture of antennas for radio astronomy.

One of the most promising new technologies is reflectarrays. The reflectarray concept is not new, but the rapid development of microstrip antenna technology, coupled with the need for low-cost and high-gain antennas for commercial applications, has led to the use of microstrip elements in a variety of reflectarray configurations. Microstrip reflectarrays combine the performance versatility of antenna arrays with the simplicity of reflectors. They are low-profile, low-mass, inexpensive, easy-to-build, passive devices with phase controlling capability. By properly adjusting the reflection phase of each sub-reflectarray element, a uniform phase distribution is achieved at the reflector aperture. The use of microstrip reflectarrays is currently largely diffused in many application fields, such as remote sensing and satellite communications. With new emerging applications, advanced features are being continuously developed in terms of broadband, multi-band, dual-polarization, and beam-scanning capabilities.

Radio astronomy has so far seen no application of reflectarrays. Besides to the relative newness of this technology, the lack of applications of reflectarrays to the design of radio telescopes is undoubtedly due to the specific requirements in this field, which usually includes large collecting areas (compared to applications in remote sensing and satellite communications), multi-band and broadband operations, and fully-steerable antennas. In this work, we have seen that reflectarrays can be successfully used from microwave to millimeter wave frequencies, and they also offer the possibility to perform the pointing and tracking operations electronically, which is of great interest for radio astronomical applications. However, we have also shown that their current limitations in terms of maximum size, bandwidth, and beam-scanning range may restrict their application to the design of radio telescopes.

The examples described in this chapter demonstrate the results that can be obtained from microstrip reflectarrays, even with their current limitations. Given the great variety of radio astronomical projects, and the telescopes designed for specific applications, it is not possible to give a unique recipe for the use of reflectarrays in the design of radio telescopes. However, it would be desirable that for any future project careful engineering studies will be carried out in order to determine which design and fabrication methods, with or without reflectarrays, represent the optimum trade-off given the required scientific goals, telescope performance matrix, and of course the projected final cost.

Abbreviations

ALMA	Atacama large millimeter array
AtLAST	Atacama large aperture submillimeter telescope
CCAT	Cerro Chajnantor Atacama telescope
EM	electromagnetic
FOV	field of view
FSS	frequency-selective surfaces
FYST	Fred young submillimeter telescope
GBT	green bank telescope
GrIn	gradient index
ILMT	international liquid mirror telescope
IRS	intelligent reflecting surfaces
LBR	large balloon reflector
LMT	large millimeter telescope
ngVLA	new generation very large array
OAM	orbital angular momentum
PAF	phased array feed
PSS	phase-shifting surfaces
SKA	square kilometer array
SPRA	spherical-phase reflectarray antenna
SPT	south pole telescope
SRT	Sardinia radio telescope

Author details

Luca Olmi
INAF, Osservatorio Astrofisico di Arcetri, Italy

*Address all correspondence to: luca.olmi@inaf.it

References

[1] Olmi L. The next generation of large millimeter- and submillimeter-wave radio telescopes. In: Recent Res. Devel. Astronomy & Astrophysics. Trivandrum, India: Research Signpost; 2003

[2] Chippendale AP, Colegate TM, O'Sullivan JD. SKAcost: A tool for SKA cost and performance estimation. SKA Telescope, Memo. 2007;**92**:1-65

[3] Ali A, Mitra A, Aissa B. Metamaterials and Metasurfaces: A review from the perspectives of materials. Mechanisms and Advanced Metadevices, Nanomaterials. 2022;**12**:1027. DOI: 10.3390/nano12061027

[4] Gagnon N, Petosa A, McNamara DA. Research and Development on phase-shifting surfaces (PSSs). IEEE Antennas and Propagation Magazine. 2013;**55**(2): 29-48

[5] Li A, Singh S, Sievenpiper D. Metasurfaces and their applications. Nanophotonics. 2018;7(6):989-1011. DOI: 10.1515/nanoph-2017-0120

[6] Roederer AG. Reflectarray Antennas. In: The 3rd European Conference on Antennas and Propagation. Berlin, Germany: IEEE Xplore; 2009

[7] Carrasco E, Encinar JA. Reflectarray antennas: A review. Proceedings of the Forum for Electromagnetic Research Methods and Application Technologies (FERMAT). 2016

[8] Khayatian B, Rahmat-Samii Y, Huang J. Radiation chararcteristics of reflectarray antennas: Methodology and applicatios to dual configurations. In: Proc. EuCAP 2006, Nice, France: IEEE Xplore; 2006

[9] Nayeri P, Yang F, Elsherbeni AZ. Beam-scanning reflectarray antennas. IEEE Antennas and Propagation Magazine. 2015;**57**(4): 32-47. DOI: 10.1109/MAP.2015. 2453883

[10] Roederer. A. Reflector Antenna Comprising a Plurality of Panels. US Patent, US6411255; 2002

[11] Nayeri P, Yang F, Elsherbeni AZ. Studies on planar Reflectarrays with spherical-phase distribution for 2-D beam-scanning. In: IEEE Antennas and Propagation Society International Symposium. Florida, U.S.A.; 2013

[12] Xu S, Rahmat-Samii Y. A compensated spherical reflector antenna using sub-reflectarrays. Microwave and Optical Technology Letters. 2009; **51**(2):577-582. DOI: 10.1002/mop.24088

[13] Costanzo S, Venneri F, Di Massa G, Borgia A, Raffo A. Bandwidth performances of reconfigurable Reflectarrays: State of art and future challenges. Radioengineering. 2018;**27** (1):1-9

[14] Bayat A, Niyaraki VS. Improvement in Reflectarray antenna bandwidth with changing the geometrical shape. Open Journal of Antennas and Propagation. 2014;**2**:21-28. DOI: 10.4236/ ojapr.2014.22003

[15] Nayeri P, Yang F, Elsherbeni AZ. Reflectarray Antennas: Theory, Designs, and Applications. Hoboken, New Jersey, U.S: Wiley; 2018

[16] Venneri F, Costanzo S, Di Massa G. Bandwidth behavior of closely spaced aperture-coupled Reflectarrays. International Journal of Antennas and Propagation. 2012;**2012**:846017. DOI: 10.1155/2012/846017

[17] Ali A, Khalily M, Brown T, Tafazolli R. Metasurface-based THz reflectarray antenna with vortex multiplexing and beam-steering capabilities for future wireless communications. iScience. 2022; **25**:104725. DOI: 10.1016/j.isci.2022.104725

[18] Oh S-W, Lee J-K, Huang J, Chang K. A six-band reflectarray antenna. In: 2009 IEEE Antennas and Propagation Society International Symposium, North Charleston, SC, USA: IEEE Xplore; 2009. DOI: 10.1109/APS.2009.5171592

[19] Guclu C, Perruisseau-Carrier J, Aydin Civi O. Proof of concept of a dual-band circularly-polarized RF MEMS beam-switching Reflectarray. IEEE Transactions on Antennas and Propagation. 2012;**60**(11):5451-5455. DOI: 10.1109/TAP.2012.2207690

[20] Rodrigo D, Jofre L, Perruisseau-Carrier J. Unit cell for frequency-tunable Beamscanning Reflectarrays. IEEE Transactions on Antennas and Propagation. 2013;**61**(12):5992-5999. DOI: 10.1109/TAP.2013.2281375

[21] Niccolai A, Grimaccia F, Mussetta M, Zich R, Gandelli A. Optimization environment definition for beam steering Reflectarray antenna design. Mathematics. 2021;**10**:33. DOI: 10.3390/math10010033

[22] Reis JR, Vala M, Caldeirinha RFS. Review paper on Transmitarray antennas. IEEE Access. 2019;7:94171-94188. DOI: 10.1109/ACCESS.2019.2924293

[23] Ali Q, Shahzad W, Ahmad I, Safiq S, Bin X, Muzahir Abbas S, et al. Recent developments and challenges on beam steering characteristics of reconfigurable Transmitarray antennas. Electronics. 2022;**11**:587. DOI: 10.3390/electronics11040587

[24] Arrebola M, Hu W, Encinar JA, Cahill R, Fusco V, Gamble HS, et al. 94 GHz beam scanning dual-reflector antenna with a sub-reflectarray. In: 30th ESA Antenna Workshop on Antennas for Earth Observation, Science, Telecommunication and Navigation Space Missions. Noordwijk (The Netherlands); 2008

[25] Rajagopalan H, Xu S, Rahmat-Samii Y. Experimental demonstration of Reflectarrays acting as conic section subreflectors in a dual reflector system. IEEE Transactions on Antennas and Propagation. 2013;**61**(11):5475-5484. DOI: 10.1109/TAP.2013.2277716

[26] Tamminen A, Gallacher TF, Ala-Laurinaho J, Räisänen AV. Developments of a reflectarray and its element characterization at millimeter wavelengths. In: Proc. SPIE 9078, Passive and Active Millimeter-Wave Imaging XVII. 2014. p. 907807. DOI: 10.1117/12.2053308

[27] Tienda C, Arrebola M, Encinar JA. Recent developments of Reflectarray antennas in dual-reflector configurations. International Journal of Antennas and Propagation. 2012;**2012**: 125287. DOI: 10.1155/2012/125287

[28] Tienda C, Encinar JA, Carrasco E, Arrebola M. Design of Dual-Reflectarray Antenna for beam scanning. Journal of Wireless Networking and Communications. 2012;**2**(1):9-14. DOI: 10.5923/j.jwnc.20120201.03

[29] Yoon JH, Yoon YJ, Lee W, So J. Axially symmetric dual-reflectarray antennas. Electronics Letters. 2014;**50**(13):908-910

[30] Arrebola M, Carrasco E, Encinar JA. Beam scanning antenna using a Reflectarray as sub-reflector. Aces Journal. 2011;**26**(6):1-11

[31] Olmi L. Reflectarrays in radio telescopes design: A preliminary analysis. In preparation

[32] Yang F, Nayeri P, Elsherbeni AZ. Recent advances in beam-scanning Reflectarray antennas. In: In XXXIth URSI GASS. Beijing, China: IEEE Xplore; 2014. DOI: 10.1109/URSIGASS.2014.6929142

[33] Olmi L, Nesti R. A preliminary study for a compensated large spherical reflector antenna using sub-reflectarrays. Experimental Astronomy. submitted

[34] Lau JY, Hum SV. Reconfigurable Transmitarray design approaches for beamforming applications. IEEE Transactions on Antennas and Propagation. 2012;**60**(12):5679-5689. DOI: 10.1109/TAP.2012.2213054

[35] Savini G, Shitvov A, Hargrave P, Ade P, Tucker C, Walker I, et al. MetaTel: Ongoing work on a meta-material sub-THz telescope for earth observing. In: 14th UK-Europe-China Workshop on Millimetre-Waves and Terahertz Technologies (UCMMT). Lancaster, United Kingdom: IEEE Xplore; 2021. DOI: 10.1109/UCMMT53364.2021.9569880

Chapter 7

The Latest Development of Tunable Microwave Filters

Zhongmao Li, Mengjie Qin, Lanlan Liao, Xiaozhuang Hu, Pengzhan Liu, Xin Qiu and Zhiqiang Li

Abstract

There have been many types of filters for various communication implications such as surface acoustic wave (SAW) filters, bulk wave (BAW) filters, transmission line filters, cavity filters, substrate integrated waveguide (SIW) filters, etc. As the radio frequency (RF) front end continues to develop, bandpass filters (BPFs) with excellent frequency response and tunability are becoming more and more competitive. In this chapter, three different types are reported, including On-chip-based TBPF, microstrip-based TBPF, and SIW-based TBPF, and some of them are theoretically analyzed carefully. Filters with frequency tuning, bandwidth tuning, and transmission zeros tuning are shown comprehensively. In addition, the limitations and future trends of tunable microwave filters are objectively expounded.

Keywords: tunable bandpass filter, microstrip filter, on-chip-based filter, SIW filter, microwave filter, radio-frequency front-end

1. Introduction

In recent years, to adapt to the technical requirements of wireless communication, the RF front-end is developing towards multi-band, tunable, miniaturized, low-cost etc. As a key frequency-selective component in the RF front-end, tunable filters increase the range of frequency and reduce the complexity in multi-band communication systems. The performance of a bandpass filter has attracted more attention. Electronically tuned bandpass filters are the most promising frequency selection scheme for tunable RF front end. Filters with different structure are introduced in the order of frequency tuning, frequency and bandwidth tuning, transmission zero tuning, and band pass filter with special function. This chapter introduces chip TBPF, micro-strip line TBPF, and substrate-integrated waveguide TBPF based on the form of tunable bandpass filters.

2. On-Chip-based TBPF

For microwave frequencies, a varactor-tuned LC filter is a popular choice in tunable BPF circuit designs, though its limited capacitance tuning range is still a problem.

IntechOpen

In 2016, a 5/60 GHz CMOS receiver with a transformer-coupled Q-enhanced tunable channel-select active filter was proposed in [1]. The partial diagram of the circuit for the 5 GHz receiver path is shown in **Figure 1**. Two varactors, C_{t1} and C_{t2} (V_T control), tune frequencies from 5.15 to 5.65 GHz. The Q factor is compensated by tuning the tail current (V_{q1} control) to get a good frequency response. By adjusting the frequency and the Q value at the same time, the filter maintains a good frequency response in different operating frequencies. In 2017, a similar receiver architecture using a negative resistance to enhance Q achieved a wider tuning range [2].

In [3], a passive LC filter is demonstrated, which consists of a 3rd-order pi-section structure to meet the out-of-band attenuation requirement. The filter provides a 300–882 MHz operating frequency range using metal-insulator-metal (MIM) capacitors and NMOS switches. In [4], a 0.8/2.4 GHz tunable active-RC BPF is reported. An elliptic filter structure is adopted to realize a large out-of-band rejection. An operational amplifier (OPAMP) offers enough gain to maintain a high Q factor. The tunability of this BPF is achieved by resistor and capacitor banks.

Nicolo Testi et al. have fabricated a 4th-order tunable Q-enhanced BPF [5]. **Figure 2** shows the complete structure of the 4th-order BPF and each stage. As shown in **Figure 2(b)**, an array of differential pairs on the left is used to provide an overall gain in the 72 dB range. The switched capacitor array on the right is used to tune the center frequency. In addition, the negative resistance network (a cross-coupled pair M6, M7) compensates for the loss of the LC tank during frequency tuning. The filter covers a frequency response from 2.35 to 2.48GHz and the frequency step is 0.4 MHz.

An N-path structure is equivalent to a series LC coupling network is proposed in [6]. As is shown in **Figure 3**, quarter-wave T-lines are added on either side to transform a shunt LC into a series LC. This structure completes the N-path equivalence theory. Moreover, T-lines also eliminate the influence between each stage of an N-path filter. **Figure 4** shows the diagram of the proposed 6th-order filter. The quarter-wave T-lines are realized by lumped capacitor-inductor-capacitor T-type circuits. The designed filter covers the 600-850 MHz tuning range. The BPF has excellent filter shape and OOB rejection but increases in-band loss.

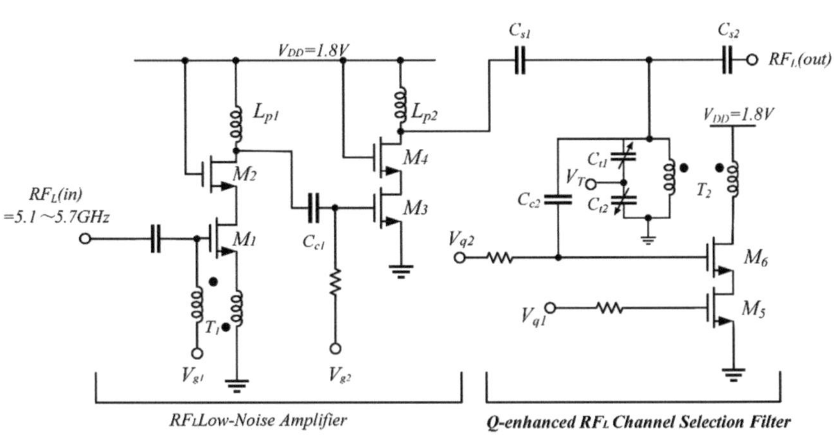

Figure 1.
Partial diagram of the circuit for the 5GHz mode in [1].

DOI: http://dx.doi.org/10.5772/intechopen.1002387

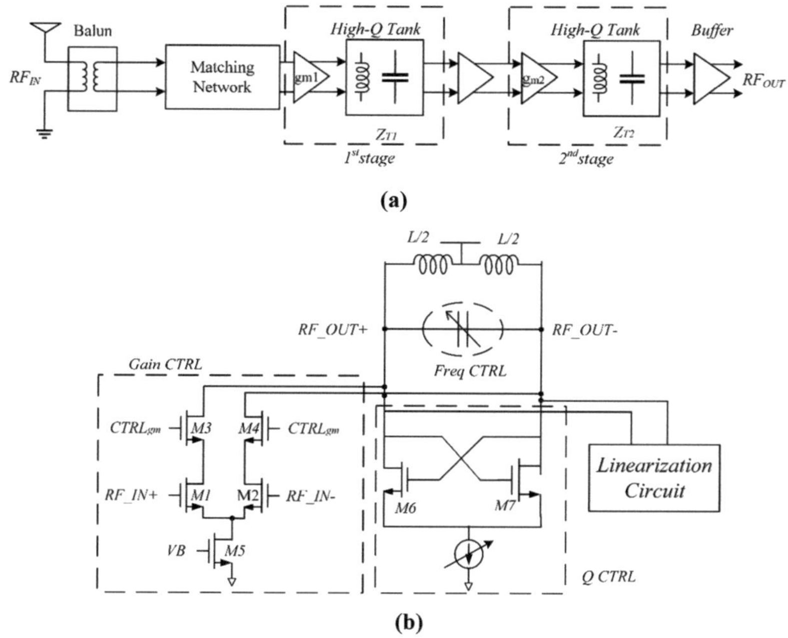

Figure 2.
Schematic of (a) the 4th-order LC TBPF and (b) each of the TBPF stages of the BPF in [5].

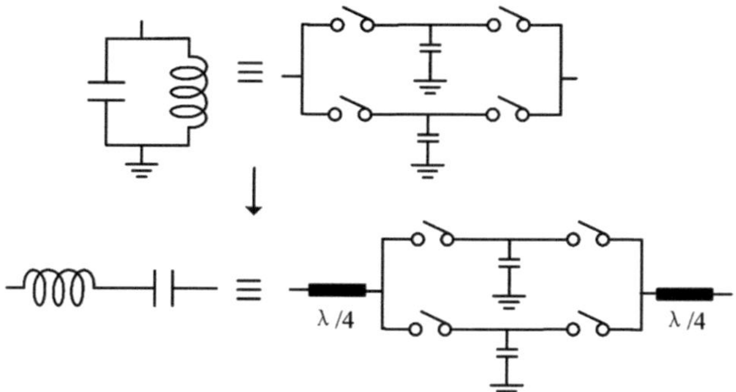

Figure 3.
Conversion for parallel LC and series LC to N paths.

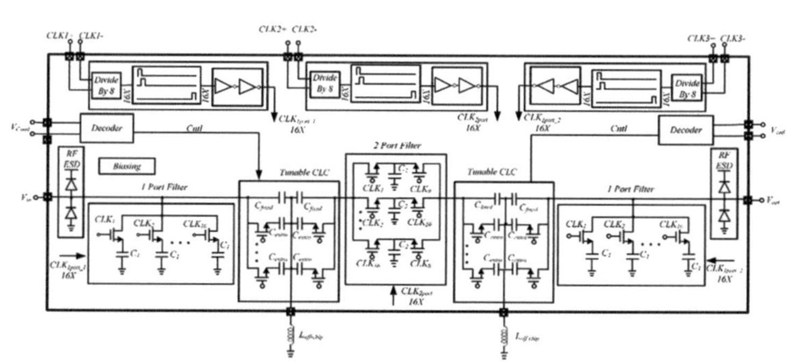

Figure 4.
Diagram of the designed 6th-order N-path filter in [6].

3. Microstrip-based TBPF

On-chip filter has the advantage of a small area and easy integration with the entire RF front-end, but the design flexibility is greatly reduced and limited by process. In addition, its performance is poor, and cost is higher. Planar BPFs, such as microstrip filters and substrate-integrated waveguide (SIW) filters, are a more popular choice due to their combination of performance, design flexibility, and ease of integration with other active elements.

In [7], a method that uses a transmission line, stub, and capacitor to form a basic resonator was proposed. An Nth-order filter can be designed by cascading multiple resonators. And it controls all varactors by a single bias, which simplifies the control circuit. In [8], a 4th-order band-pass filter was designed using two loop hairpin resonators and two combined resonators. This structure has a more symmetrical in-band response. In [9], the best impedance- to length ratio of the stepped-impedance resonator (SIR) was demonstrated. Fabricated SIR-loaded BPF achieved the widest bandwidth tuning. In [10], a novel filter using comb line and split-ring resonators achieved a tuning range of 1.1–1.88 GHz.

One way to implement a BPF with tunable bandpass and frequency is to cascade two filter units with different frequency responses. In [11], a BPF with an adjustable pass-band edge was designed by cascading a high-pass filter (HPF) and a low-pass filter (LPF) directly. **Figure 5** shows the layout of proposed filter. HPF is on the left and LPF is on the right. The passband edges are the cutoff frequency of LPF and HPF controlled by loaded varactors. Its tuning range is determined by capacitance of varactor and the electric length of the coupled line. The fabricated filter has a sharp roll-off and good in-band shape.

Another way to vary bandwidth is to adjust coupling, including interstage coupling and input-output coupling. A tunable filter based on cross-shaped multi-mode resonators are shown in **Figure 6** [12]. The input/output coupling circuits (J/K inverter cascaded with a transformer) control the resonator to operate at different mode frequencies. The odd mode is controlled by varactor C_1 while the even mode and

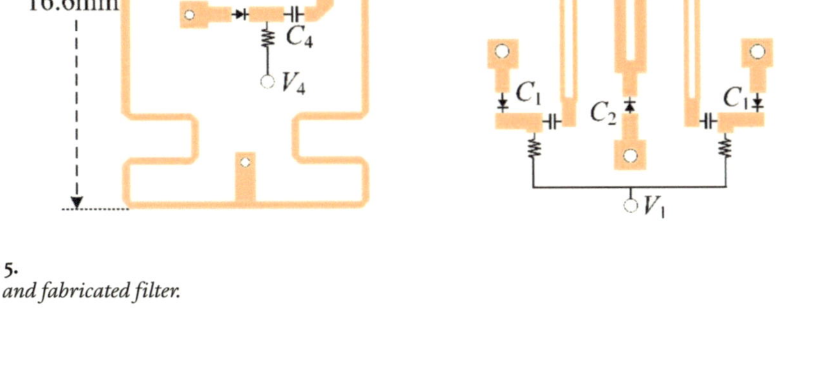

Figure 5.
Layout and fabricated filter.

DOI: http://dx.doi.org/10.5772/intechopen.1002387

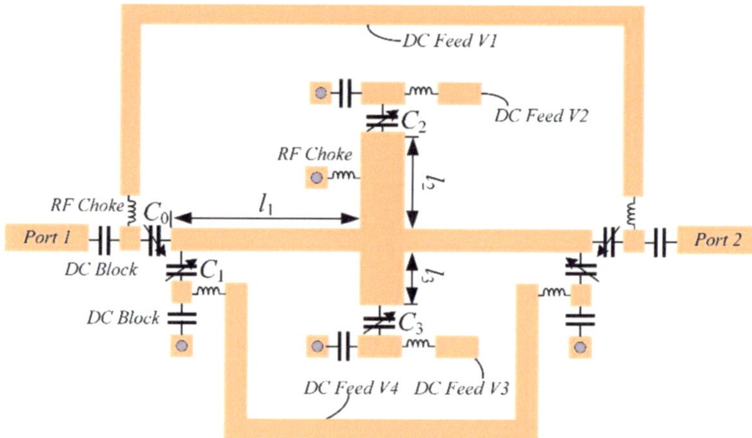

Figure 6.
Configuration of the proposed filter.

transmission zeros are controlled by C_2 and C_3. The center frequency of the fabricated filter can be tuned from 600 to 1450 MHz. The return loss is more than 15 dB and insertion loss is less than 4 dB over tuning range. The Bandwidth can vary from 120 to 950 MHz when center frequency is 1 GHz.

In ultra-wideband (UWB) system, many narrowband interference signals affect the information reception. Thus, a notch filter is needed to protect the system from unwanted interference. Notch bandpass filter that combines pre-selection and interference suppression is a multi-functional filter. Existing notch design schemes mainly include using additional notch resonators [13], embedded open stubs [14], and asymmetric coupled fed lines [15]. To suppress in-band interference dynamically, notch tunable filters are required [16–18]. Recently, reconfigurable notch BPF with tunable bandwidth has also been reported [19–22].

Based on a pair of parallel-coupled lines (PCLs) and a terminated cross-shaped resonator (TCSR), a series of wideband reconfigurable BPFs with notch control are designed in [19], including tunable notch frequency, tunable notch bandwidth, and tunable broadband bandwidth. Each part can be controlled independently, which greatly improves the tuning freedom of notch bandpass filters. **Figure 7** illustrates filter A. The notch bandwidth is controlled by S1 and SF independently, and the notch frequency is affected by $l_{l\text{-}H}$ and $l_{l\text{-}V}$. **Figure 8** presents the filter B. Varactors are inserted to vary the parameters of TCSR. The BPF bandwidth can be tunable by the variable parameters of TCSR. Measured results present that notch frequency and bandwidth keep fixed when the BPF operate in different bandwidth. The bandwidth tuning range is 1.22–2.46 GHz.

4. SIW-based TBPF

In [23], a tunable dual-band filter is reported using half-mode substrate-integrated waveguide (SIW) structure. Two pairs of stubs are loaded at the open ends of the HMSIW structure to form the two passbands. Frequency tuning is achieved by loading variable capacitors at the end of stubs. Changes in one pair of load lines have little effect on the other passband.

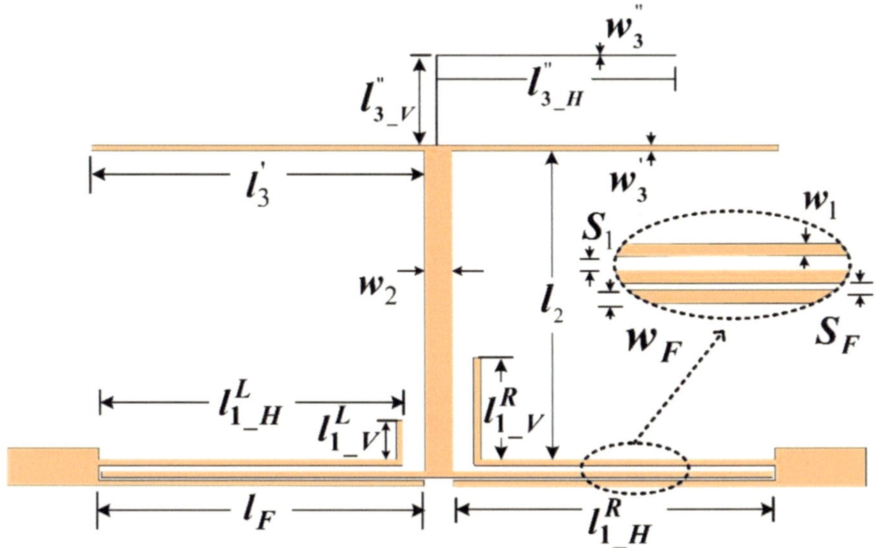

Figure 7.
Schematic of filter A.

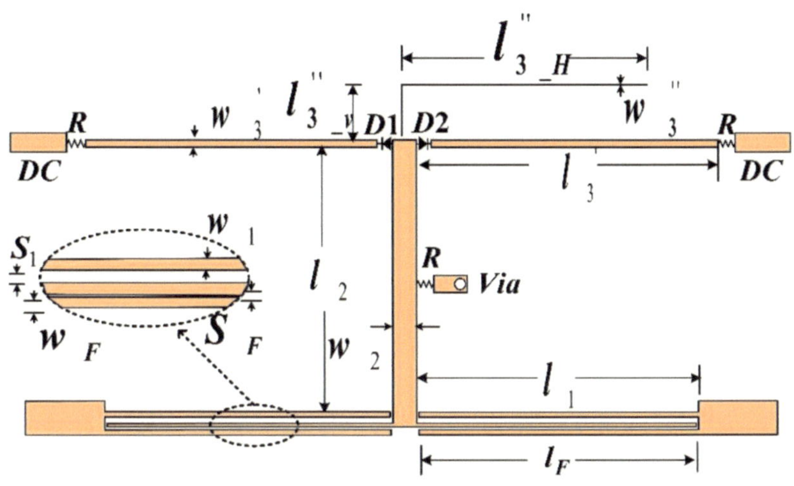

Figure 8.
Schematic of filter B.

In [24], two 4-pole SIW filters were designed with different dual-mode resonator configurations. The basic SIW cavity is excited by 50 Ω microstrip line, which loaded coupling slots at either side of transmission lines to impedance match. An H-shaped slot is introduced in the proposed rectangular cavity to generate second resonant mode. **Figure 9** shows two higher-order structures based on basic resonator. Two resonators are placed vertically or horizontally and coupled by a pair of U-shaped slots. A varactor was placed to tune frequency. The frequency tuning range of Filter A is 2.7–3.4 GHz, and filter B is 3.1–3.9 GHz. The IL of both of them is less than 1.5 dB.

DOI: http://dx.doi.org/10.5772/intechopen.1002387

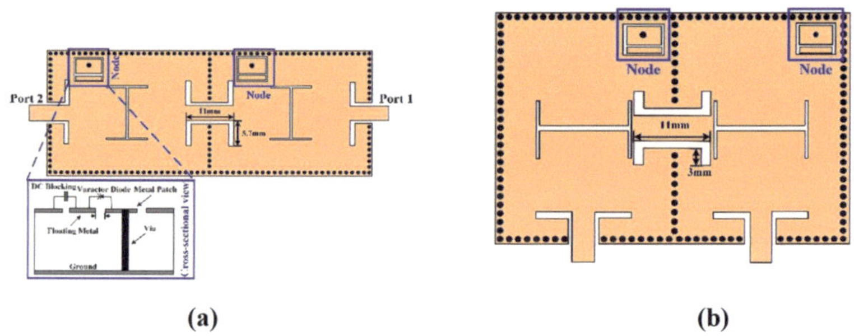

Figure 9.
Proposed filter in (a) vertical configuration (b) horizontal configuration.

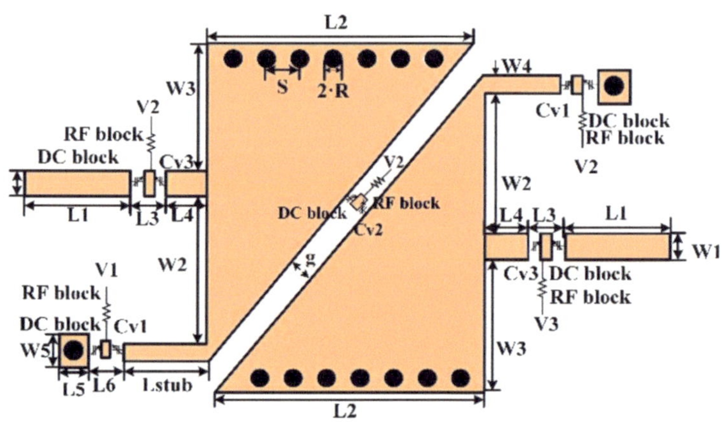

Figure 10.
Diagram of proposed SIW TBPF.

An eighth-mode SIW tunable filter was presented in [25]. As shown in **Figure 10**, varactor C_{v1} is loaded at stubs to achieve frequency tuning. Varactor C_{v2} is loaded between two resonant cavities to vary the electrical coupling strength, which enables bandwidth tuning. The frequency tuning range is 2.17–2.72 GHz and the BW tuning range is 140-435 MHz. In addition, the edge electrical coupling between two resonators offers a transmission zero out-of-band, which enables the filter to obtain better out-of-band rejection.

In [26], a half-mode SIW filter with a superposed two layers PCB was demonstrated. As shown in **Figure 11**, PCB2 is extended outside of PCB1 to load feed lines and varactors. The metal semicircular not only helps to improve the accuracy of frequency tuning but also reduces the filter size. Varactor C_{v1} is loaded between the center *via* hole to tune frequency. Varactor C_{v2} is loaded between the holes of resonator to tune the bandwidth by affecting the electric coupling and magnetic coupling. Varactor C_{v3} could provide a suitable Q factor when the frequency and bandwidth is tunable. The frequency turning range is 1.11–1.5 GHz and bandwidth is from 46 to 130 MHz.

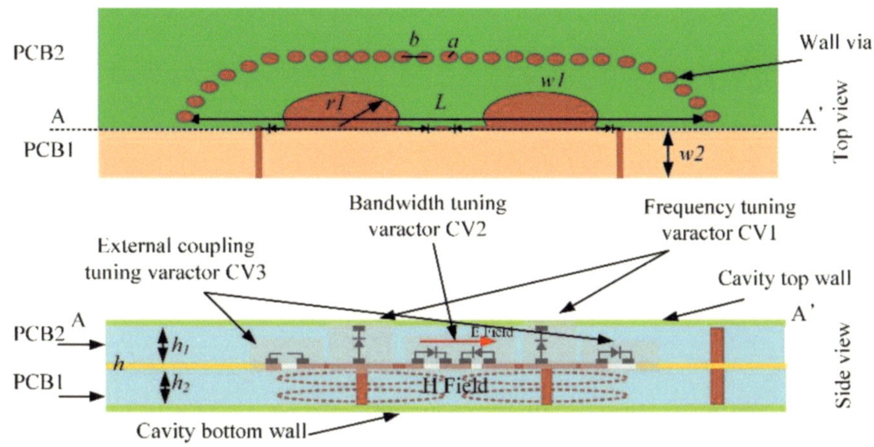

Figure 11.
Top view and the side view of proposed filter.

5. Conclusion

There is no doubt that tunable microwave filters are pushing towards more compact, more functions but less control components. In this process of development, on-chip-based TBPF, microstrip-based TBPF, and SIW-based filter are seen as three potential ways. On-chip BPF provide the possibility of full integration of system-on-chips. Microstrip line TBPF demonstrates flexible filtering performances. SIW-based filter has less loss in the band. However, the low Q value of filters in CMOS and GaAs process, the dispersion of microstrip-based filter in high frequency, and the relatively large area of SIW-based filters are still unresolved issues. Moreover, a more universal accurate design process and a simpler control method are very meaningful to research.

Author details

Zhongmao Li*, Mengjie Qin, Lanlan Liao, Xiaozhuang Hu, Pengzhan Liu, Xin Qiu and Zhiqiang Li
Institute of Microelectronics of the Chinese Academy of Sciences, Beijing, P.R. China

*Address all correspondence to: lzm225@foxmail.com

References

[1] Hsiao Y-C, Meng C, S.-T. Yang, 5/60 GHz 0.18 um CMOS dual-mode dual-conversion receiver using a Tunable active filter for 5-GHz Channel selection. IEEE Microwave and Wireless Components Letters. 2016;**26**(11):951-953. DOI: 10.1109/LMWC.2016.2615012

[2] Chang WL, Meng C, Yang S-D, Huang G-W. 0.18 μm SiGe BiCMOS microwave/millimeter-wave dual-mode dual-conversion receiver architecture with a tunable RF channel selection at low-flicker-noise microwave mode. In: 2017 IEEE MTT-S International Microwave Symposium (IMS). Honololu, HI, USA: IEEE; Jun 2017. pp. 1778-1780. DOI: 10.1109/MWSYM.2017.8058992

[3] Im D, Kim H, Lee K. A broadband CMOS RF front-end for universal tuners supporting multi-standard terrestrial and cable broadcasts. IEEE Journal of Solid-State Circuits. 2012;**47**(2):392-406. DOI: 10.1109/JSSC.2011.2168650

[4] Xu Z et al. A 0.8/2.4 GHz Tunable active band pass filter in InP/Si BiCMOS technology. IEEE Microwave and Wireless Components Letters. 2014;**24**(1):47-49. DOI: 10.1109/LMWC.2013.2287236

[5] Testi N, Berenguer R, Zhang X, Munoz S, Xu Y. A 2.4GHz 72dB-variable-gain 100dB-DR 7.8mW 4th-order tunable Q-enhanced LC band-pass filter. In: 2015 IEEE Radio Frequency Integrated Circuits Symposium (RFIC). Phoenix, AZ, USA: IEEE; 2015. pp. 87-90. DOI: 10.1109/RFIC.2015.7337711

[6] Reiskarimian N, Krishnaswamy H. Design of all-passive higher-order CMOS N-path filters. In: 2015 IEEE Radio Frequency Integrated Circuits Symposium (RFIC). Phoenix, AZ, USA: IEEE; 2015. pp. 83-86. DOI: 10.1109/RFIC.2015.7337710

[7] Lim T, Anand A, Chen J, Liu X, Lee Y. Design method for Tunable planar Bandpass filters with single-bias control and wide Tunable frequency range. IEEE Transactions on Circuits and Systems II: Express Briefs. 2021;**68**(1):221-225. DOI: 10.1109/TCSII.2020.3004614

[8] Zakharov A, Rozenko S, Ilchenko M. Varactor-tuned microstrip Bandpass filter with loop hairpin and Combline resonators. IEEE Transactions on Circuits and Systems II: Express Briefs. 2019;**66**(6):953-957. DOI: 10.1109/TCSII.2018.2873227

[9] Qin W, Cai J, Li Y-L, Chen J-X. Wideband Tunable Bandpass filter using optimized Varactor-loaded SIRs. IEEE Microwave and Wireless Components Letters. 2017;**27**(9):812-814. DOI: 10.1109/LMWC.2017.2734848

[10] Zhao Z, Chen J, Yang L, Chen K. Three-pole Tunable filters with constant bandwidth using mixed Combline and Split-ring resonators. IEEE Microwave and Wireless Components Letters. 2014;**24**(10):671-673. DOI: 10.1109/LMWC.2014.2340996

[11] Bi X-K, Zhang X, Wong S-W, Guo S-H, Yuan T. Reconfigurable-bandwidth DWB BPF with fixed operation frequency and controllable stopband. IEEE Transactions on Circuits and Systems II: Express Briefs. 2021;**68**(1):141-145. DOI: 10.1109/TCSII.2020.3001872

[12] Mao J-R, Choi W-W, Tam K-W, Che WQ, Xue Q. Tunable Bandpass filter design based on external quality factor tuning and multiple mode resonators for wideband applications. IEEE

Transactions on Microwave Theory and Techniques. 2013;**61**(7):2574-2584. DOI: 10.1109/TMTT.2013.2264684

[13] Song Y, Yang G-M, Geyi W. Compact UWB Bandpass filter with dual notched bands using defected ground structures. IEEE Microwave and Wireless Components Letters. 2014;**24**(4):230-232. DOI: 10.1109/LMWC.2013.2296291

[14] Shaman H, Hong J-S. A novel ultra-wideband (UWB) Bandpass filter (BPF) with pairs of transmission zeroes. IEEE Microwave and Wireless Components Letters. 2007;**17**(2):121-123. DOI: 10.1109/LMWC.2006.890335

[15] Song K, Xue Q. Compact ultra-wideband (UWB) Bandpass filters with multiple notched bands. IEEE Microwave and Wireless Components Letters. 2010;**20**(8):447-449. DOI: 10.1109/LMWC.2010.2050303

[16] Kurra L, Abegaonkar MP, Basu A, Koul SK. Switchable and Tunable notch in ultra-wideband filter using electromagnetic bandgap structure. IEEE Microwave and Wireless Components Letters. 2014;**24**(12):839-841. DOI: 10.1109/LMWC.2014.2363020

[17] Wang H, Tam K-W, Ho S-K, Kang W, Wu W. Design of Ultra-Wideband Bandpass Filters with Fixed and Reconfigurable Notch Bands Using Terminated Cross-Shaped Resonators. IEEE Transactions on Microwave Theory and Techniques. 2014;**62**(2):252-265. DOI: 10.1109/TMTT.2013.2296530

[18] Wu Z, Shim Y, Rais-Zadeh M. Miniaturized UWB filters integrated with Tunable notch filters using a silicon-based integrated passive device technology. IEEE Transactions on Microwave Theory and Techniques. 2012;**60**(3):518-527. DOI: 10.1109/TMTT.2011.2178428

[19] Bi X-K, Zhang X, Wong S-W, Guo S-H, Yuan T. Design of Notched-Wideband Bandpass Filters with Reconfigurable Bandwidth Based on terminated cross-shaped resonators. IEEE Access. 2020;**8**:37416-37427. DOI: 10.1109/ACCESS.2020.2975379

[20] Teng C, Cheong P, Tam K-W. Reconfigurable wideband Bandpass filters based on dual cross-shaped resonator. In: 2019 IEEE MTT-S International Wireless Symposium (IWS). Guangzhou, China: IEEE; May 2019. pp. 1-3. DOI: 10.1109/IEEE-IWS.2019.8804159

[21] Psychogiou D, Gómez-García R, Peroulis D. RF wide-band Bandpass filter with dynamic in-band multi-interference suppression capability. IEEE Transactions on Circuits and Systems II: Express Briefs. 2018;**65**(7):898-902. DOI: 10.1109/TCSII.2017.2726145

[22] Psychogiou D, Gómez-García R, Peroulis D. Wide-passband filters with in-band tunable notches for agile multi-interference suppression in broad-band antenna systems. In: 2018 IEEE Radio and Wireless Symposium (RWS). Anaheim, CA, USA: IEEE; Jan 2018. pp. 213-216. DOI: 10.1109/RWS.2018.8304990

[23] Zhou C-X, Zhu C-M, Wu W. Tunable dual-band filter based on stub-capacitor-loaded half-mode substrate integrated waveguide. IEEE Transactions on Microwave Theory and Techniques. 2017;**65**(1):147-155. DOI: 10.1109/TMTT.2016.2613053

[24] Iqbal A, Ahmad AW, Smida A, Mallat NK. Tunable SIW Bandpass filters with improved upper stopband performance. IEEE Transactions on Circuits and Systems II: Express Briefs. 2020;**67**(7):1194-1198. DOI: 10.1109/TCSII.2019.2936495

[25] Guo J, You B, Luo GQ. A miniaturized eighth-mode substrate-integrated waveguide filter with both Tunable Center frequency and bandwidth. IEEE Microwave and Wireless Components Letters. 2019;**29**(7):450-452. DOI: 10.1109/LMWC.2019.2916780

[26] You B, Lu S, Chen L, Gu QJ. A half-mode substrate-integrated filter with Tunable Center frequency and reconfigurable bandwidth. IEEE Microwave and Wireless Components Letters. 2016;**26**(3):189-191. DOI: 10.1109/LMWC.2016.2526031

Chapter 8

Analysis of Microstrip Circuits Using the FDTD Method in Julia

Veysel Demir

Abstract

In this chapter we present a finite-difference time-domain (FDTD) code that is developed in Julia programming language customized for efficient electromagnetic simulation of microstrip circuits. Julia language has become popular in recent years for developing codes for scientific computing due to high-performance it can provide while being a dynamically typed language that supports interactive use like MATLAB or Python. The presented FDTD program is customized for faster simulation of microstrip circuits with lower memory requirements by employing an efficient form of FDTD updating equations, referred to as edge-length normalized formulation, as well as other simplifications that would make it sufficient to analyze single layer microstrip circuits and produce scattering parameters. We demonstrate how updating equations for fields, sources, and absorbing boundaries are modified to accommodate edge-length normalization, and how these equations are translated into a code in Julia. We then conclude the chapter with a presentation of some examples of circuits simulated with the presented code.

Keywords: finite difference time domain method, electromagnetic simulations, microstrip circuits, Julia, scattering parameters

1. Introduction

Modern engineering practice heavily utilizes simulation software packages for the analysis and design of systems. As such, engineers and researchers who practice in the area of electromagnetics, and radio frequency and microwave circuit design use electromagnetic analysis and simulation software tools based on various types of numerical methods that are developed based on Maxwell's equations. Among several such methods the finite-difference time-domain (FDTD) method [1–5], the finite element method (FEM), and the Method of Moments (MoM) are traditionally the most common methods to solve Maxwell's equations. Among these, arguably, the FDTD method is the most popular one due to its simplicity in formulation, expandability, and flexibility it can provide to analyze a large variety of types of problems, as well as it is parallelizability on various types of modern computer architectures to obtain results efficiently in terms of computation speed. In this chapter we demonstrate that one can easily develop a small FDTD program that is customized to analyze microstip circuits efficiently using the Julia programming language.

Numerical methods used for the analyses of electromagnetics problems are computationally demanding in terms of time and memory usage on computers if reasonably accurate results are sought. Improvements in computer hardware such as faster multi-CPU and GPU systems made it possible to run analysis of large-scale electromagnetics problems on common desktop computers. Still a numerical analysis method needs to be coded in a high-performance programming language to make full use of the capabilities of the existing hardware on a computer system. Traditionally Fortran and C/C++ have been the primary choices of programming languages for scientific computing as they are compiled languages, which converts a code directly into a machine code that the processor can execute very fast. Interpreted languages are another class of programming languages that is more flexible and convenient to use by someone who would like to develop a numerical analysis code, experiment with various features and algorithms, and conveniently display and analyze the results of simulations. However, these languages are significantly slower than the compiled ones. While there are many such languages Python and MATLAB can be named as among the popular ones for scientific computing. Julia programming language [6, 7] has been introduced to close the gap between the speed of compiled and flexibility of interpreted languages: Julia is dynamically typed, therefore can be used interactively like a scripting language, while just-in-time (JIT) compilation produces a compiled code that is claimed to run as fast as C or Fortran. Programmers familiar with MATLAB or Fortran may find it easier to learn Julia. In this chapter we will present an FDTD code developed in Julia; One can use this code as a base code to develop new features over and experiment with various aspects of the FDTD method.

2. FDTD updating equations with edge length normalization

The FDTD method is essentially simulation of electromagnetic fields in discrete time in a discrete three-dimensional space in a time-marching loop. It is based on Maxwell's curl equations in differential form

$$\nabla \times \overline{H} = \varepsilon \frac{\partial \overline{E}}{\partial t} + \sigma^e \overline{E}, \tag{1}$$

$$\nabla \times \overline{E} = -\mu \frac{\partial \overline{H}}{\partial t}, \tag{2}$$

where ε is permittivity, μ is permeability, and σ^e is the electric conductivity of a medium. These vector equations can be decomposed into three scalar equations each in Cartesian coordinates. For instance, one of these equations that can be obtained from (1) is

$$\varepsilon_x \frac{\partial E_x}{\partial t} + \sigma_x^e E_x = \frac{\partial H_z}{\partial y} - \frac{\partial H_y}{\partial z}, \tag{3}$$

where ε_x and σ_x^e are material parameters associate with E_x. As nonmagnetic materials and substrates are used in most microstrip circuits, we can assume that the magnetic permeability in (2) has a constant value as free space permeability μ_0. As a result, one can extract three scalar equations from (2) also, one of which is

DOI: http://dx.doi.org/10.5772/intechopen.1004059

$$\mu_0 \frac{\partial H_x}{\partial t} = \frac{\partial E_y}{\partial z} - \frac{\partial E_z}{\partial y} \tag{4}$$

In FDTD method a three-dimensional space is represented by cuboid cells, referred to as Yee cells, and electric and magnetic field components are defined at discrete positions on each of these cells. These field components are also defined in discrete time instances such that electric fields are at integer time steps (such as $0, 1, 2, \ldots, n-1, n, n+1, \ldots$) and magnetic fields are at half-integer time steps (such as $0.5, 1.5, 2.5, \ldots, n-0.5, n+0.5, \ldots$), where n indicates the indices of discrete time instants separated by a time step of Δt. The time and space derivatives in (3) and (4) are then imposed on the discrete time and space using central difference approximations to obtain equations in discrete form and these equations are arranged such that future field terms can be calculated using current and past field terms. For instance, following the notations in Elsherbeni and Demir [3], one can obtain the following equation from (3)

$$\begin{aligned} E_x^{n+1}(i,j,k) = {} & C_{exe}(i,j,k) \times E_x^n(i,j,k) \\ & + C_{exhz}(i,j,k) \times \left(H_z^{n+\frac{1}{2}}(i,j,k) - H_z^{n+\frac{1}{2}}(i,j-1,k) \right) \\ & + C_{exhy}(i,j,k) \times \left(H_y^{n+\frac{1}{2}}(i,j,k) - H_y^{n+\frac{1}{2}}(i,j,k-1) \right), \end{aligned} \tag{5}$$

where

$$C_{exe}(i,j,k) = \frac{2\varepsilon_x(i,j,k) - \Delta t \sigma_x^e(i,j,k)}{2\varepsilon_x(i,j,k) + \Delta t \sigma_x^e(i,j,k)}, \tag{6}$$

$$C_{exhz}(i,j,k) = \frac{1}{\Delta y} \frac{2\Delta t}{2\varepsilon_x(i,j,k) + \Delta t \sigma_x^e(i,j,k)}, \tag{7}$$

$$C_{exhy}(i,j,k) = -\frac{1}{\Delta z} \frac{2\Delta t}{2\varepsilon_x(i,j,k) + \Delta t \sigma_x^e(i,j,k)}. \tag{8}$$

Similarly, one can obtain the following equation from (4)

$$\begin{aligned} H_x^{n+0.5}(i,j,k) = {} & H_x^{n-0.5}(i,j,k) \\ & + \frac{\Delta t}{\Delta z \mu_0} \left(E_y^{n+0.5}(i,j,k+1) - E_y^{n-0.5}(i,j,k) \right) \\ & - \frac{\Delta t}{\Delta y \mu_0} \left(E_z^{n+0.5}(i,j+1,k) - E_z^{n-0.5}(i,j,k) \right). \end{aligned} \tag{9}$$

Such equations as (5) and (9) are called updating equations as they are used to update electric and magnetic fields distributed in a three-dimensional space at each time step of a iterative time-marching algorithm. The terms such as $C_{exe}(i,j,k)$, $C_{exhz}(i,j,k)$, and $C_{exhy}(i,j,k)$ are called updating coefficients.

When coding the FDTD algorithm, each of the field terms E_x, E_y, E_z, H_x, H_y, and H_z are stored as three-dimensional arrays in computer memory. The updating coefficient terms are also stored as three-dimensional arrays in computer memory. Notice that since we assumed all materials in the problem space are nonmagnetic, we used free space permeability μ_0 in the entire problem space, thus the updating coefficients in (9) simplified to become scalar constants. Hence, we eliminated the need to store three-dimensional arrays for updating coefficients for magnetic field updates, which reduces the memory requirement to store updating coefficients by half. We can consider this as a customization of the general FDTD formulation for the special case of microstrip circuits.

Notice that in (5) the coefficients $C_{exhz}(i,j,k)$ and $C_{exhy}(i,j,k)$ are different from each other only by factors of $1/\Delta y$ and $1/\Delta z$, and a negative sign. We can express these two coefficients as

$C_{exhz}(i,j,k) = \frac{1}{\Delta y} C_{exh}(i,j,k)$ and $C_{exhy}(i,j,k) = -\frac{1}{\Delta z} C_{exh}(i,j,k)$,

where $C_{exh}(i,j,k) = \frac{2\Delta t}{2\varepsilon_x(i,j,k) + \Delta t \sigma_x^e(i,j,k)}$.

Then instead of storing three coefficient arrays (C_{exe}, C_{exhz}, and C_{exhy}) on computer memory, we can just store two arrays (C_{exe}, C_{exh}) and multiply the terms $1/\Delta y$ and $-1/\Delta z$ with C_{exh} on the fly in each iteration. This saves memory requirement, however with the expense of two extra multiplication operations per electric field component update.

2.1 Edge length normalization

Another approach to reduce the number of coefficient arrays, however without the expense of increased number of arithmetic operations, is the method called edge length normalization [5]. Let us rewrite (5) in terms of C_{exh} as

$$\begin{aligned} E_x^{n+1}(i,j,k) = {} & C_{exe}(i,j,k) \times E_x^n(i,j,k) \\ & + \frac{1}{\Delta y} C_{exh}(i,j,k) \times \left(H_z^{n+\frac{1}{2}}(i,j,k) - H_z^{n+\frac{1}{2}}(i,j-1,k) \right) \\ & - \frac{1}{\Delta z} C_{exh}(i,j,k) \times \left(H_y^{n+\frac{1}{2}}(i,j,k) - H_y^{n+\frac{1}{2}}(i,j,k-1) \right). \end{aligned} \tag{10}$$

If we multiply both sides of the equation by Δx, multiply the second line by $\Delta z/\Delta z$, and the third line by $\Delta y/\Delta y$, we get

$$\begin{aligned} \Delta x E_x^{n+1}(i,j,k) = {} & \Delta x C_{exe}(i,j,k) \times E_x^n(i,j,k) \\ & + \frac{\Delta x \, \Delta z}{\Delta y \, \Delta z} C_{exh}(i,j,k) \times \left(H_z^{n+\frac{1}{2}}(i,j,k) - H_z^{n+\frac{1}{2}}(i,j-1,k) \right) \\ & - \frac{\Delta x \, \Delta y}{\Delta z \, \Delta y} C_{exh}(i,j,k) \times \left(H_y^{n+\frac{1}{2}}(i,j,k) - H_y^{n+\frac{1}{2}}(i,j,k-1) \right) \end{aligned} \tag{11}$$

In this equation we can normalize the field terms with lengths of the cell edges they are associated with such that e_x substitutes $\Delta x E_x$, h_y substitutes $\Delta y H_y$, and h_z substitutes $\Delta z H_z$. After these substitutions we get

$$\begin{aligned} e_x^{n+1}(i,j,k) = {} & C_{exe}(i,j,k) \times e_x^n(i,j,k) \\ & + \frac{\Delta x}{\Delta y \Delta z} C_{exh}(i,j,k) \times \left(h_z^{n+\frac{1}{2}}(i,j,k) - h_z^{n+\frac{1}{2}}(i,j-1,k) \right) \\ & - \frac{\Delta x}{\Delta y \Delta z} C_{exh}(i,j,k) \times \left(h_y^{n+\frac{1}{2}}(i,j,k) - h_y^{n+\frac{1}{2}}(i,j,k-1) \right). \end{aligned} \tag{12}$$

Now we can merge $\frac{\Delta x}{\Delta y \Delta z}$ into C_{exh} and rewrite (12) as

DOI: http://dx.doi.org/10.5772/intechopen.1004059

$$e_x^{n+1}(i,j,k) = C_{exe}(i,j,k) \times e_x^n(i,j,k) + C_{exh}(i,j,k) \times \begin{pmatrix} h_z^{n+\frac{1}{2}}(i,j,k) - h_z^{n+\frac{1}{2}}(i,j-1,k) \\ -h_y^{n+\frac{1}{2}}(i,j,k) + h_y^{n+\frac{1}{2}}(i,j,k-1) \end{pmatrix}, \tag{13}$$

where $C_{exh}(i,j,k) = \frac{\Delta x}{\Delta y \Delta z} \frac{2\Delta t}{2\varepsilon_x(i,j,k)+\Delta t \sigma_x^e(i,j,k)}$.

Eq. (13) is referred to as edge length normalized formulation. This formulation helps to reduce the total number of arithmetic operations while keeping the number of three-dimensional coefficient arrays at two.

Similarly, edge length normalization can be applied to (9) to obtain equations for magnetic field updates also as

$$h_x^{n+0.5}(i,j,k) = h_x^{n-0.5}(i,j,k) + C_{hxe} \begin{pmatrix} e_y^{n+0.5}(i,j,k+1) - e_y^{n-0.5}(i,j,k) \\ -e_z^{n+0.5}(i,j+1,k) + e_z^{n-0.5}(i,j,k) \end{pmatrix}, \tag{14}$$

where $C_{hxe} = \frac{\Delta x}{\Delta y \Delta z} \frac{\Delta t}{\mu_0}$.

Since the normalized values of fields are computed instead of the actual values of fields at every time step, formulations of all other computations also need to be adjusted to accommodate this edge length normalization. Such computations include updating voltage sources, updating lumped element components, updating absorbing boundaries, and capturing voltages and currents.

2.2 Absorbing boundaries

We simulate electric and magnetic fields in a rectangular three-dimensional domain with finite size. The boundaries of this domain need to be treated according to the type of simulation sought. For instance, a microwave circuit can be simulated in a closed box boundaries of which are treated as perfect electric conductor (PEC). Scattering problems or antenna radiation problems need to be simulated in open space. In that case an open space is simulated by treating the boundaries of the FDTD domain as absorbing boundaries. Several different types of absorbing boundary formulations have been proposed in the literature, one type of which is referred to as perfectly matched layers (PML). In this implementation we use a variation of the Auxiliary Differential Equation Perfectly Matched Layers (ADE-PML) [8] to terminate five sides of the boundaries of the FDTD domain in which a microstrip circuit is simulated, and one boundary treated as PEC to simulate the ground of the microstrip circuit.

While ADE-PML is a very effective formulation to suppress reflections from the boundaries of a problem space, it is computationally expensive, especially for a microstrip circuit simulation, as PML regions will cover a significant portion of the three-dimensional problem space. One can consider a lightweight absorbing boundary formulation, such as Liao [9] or Mur [10], to terminate the boundaries of the problem space. While such formulations are not as efficient in suppressing reflections, they are more efficient computationally, and they can provide sufficient accuracy for the scattering parameters of microstrip circuits in shorter time.

It should be noted that if it is desired to simulate a microstrip circuit in a PEC box, the ADE-PML computations can be turned off and the program will simulate PEC boundaries by default in the presented FDTD implementation. However, one should notice that when a circuit is simulated in a closed box, electromagnetic fields bounce back and forth between the walls of the box, leading to box resonances, and it will take very long for the transients vanish; therefore, the FDTD simulations of such circuits will take very long time.

3. Programming FDTD in Julia

In this section we discuss the Julia implementation of the FDTD procedure that is customized for simulation of microstrip circuits conveniently; One can easily set up a single layer multi-port microstrip circuit, run an analysis, and obtain the scattering parameters of the circuit. While the implementation is specific to microstrip simulations, it is easy to use this implementation as a base code and expand on it to build a more general purpose FDTD program that can be used for other types of applications such as scattering from complex objects, or radiation from antennas. The source code of the presented implementation can be accessed at [11].

Listing 1 shows the function, defined as "run_fdtd()", that can be called in Julia terminal to start an FDTD simulation. This function needs to be called with a project file name. A second argument, which is optional, can be used to run the simulation if it is "run_simulation" or can be used to just display the three-dimensional geometry of the problem space and material distribution if it is "display_problem_space". Several steps of an FDTD simulation process are executed by respective function calls in this function. These steps are mainly loading the project from a file, initializing the problem space, displaying the problem space, initializing the updating coefficient and field arrays, running the time marching loop, calculating scattering parameters and displaying the simulation results.

Listing 1. The function to run the FDTD simulation.

```
function run_fdtd(project_file_name, v::String = "run_simulation")
    println("===<Project name is $project_file_name >=====")
    fdtd_project = load_project_from_file(project_file_name);
    fdtd_domain, material_grid = initialize_problem_space(fdtd_project);
    display_problem_space(fdtd_project, fdtd_domain, material_grid)
    if (v != "run_simulation")
        return nothing
    end
    fdtd_fields, fdtd_coefficients, pml_boundary =
    initialize_updating_coefficients_and_fields(fdtd_project,
                    fdtd_domain, material_grid);
    run_time_marching_loop(fdtd_project, fdtd_domain, fdtd_fields,
                    fdtd_coefficients, pml_boundary)
    calculate_scattering_parameters(fdtd_project)
    display_simulation_results(fdtd_project)
    return nothing
end
```

The input project file to the "run_fdtd()" function is an Extensible Markup Language (XML) file. XML is a well-known file format used for storing and transmitting data. It is a simple human-readable file format and open-source packages are available in many programming languages, including Julia, to read and process XML files. "XMLDict" is one of such packages and used in the presented code.

DOI: http://dx.doi.org/10.5772/intechopen.1004059

All the parameters that are necessary to run the simulation of a single layer microstrip circuit need to be defined in and are read from the XML project file. The file includes analysis parameters such as cell sizes in x, y, and z directions and maximum number of time steps to run the simulation. It includes the relative permittivity, electric conductivity, thickness, and coordinates of the substrate. One can enter the geometry of the microstrip structures using rectangular bricks or polygonal shapes that are on the surface of the substrate. Multiple ports can be defined by their coordinates and impedances. Ports extend from ground to the surface of the substrate. Vias can be defined with coordinates of rectangular regions extending from the ground to the surface of the substrate. Moreover, one can define the frequency range for which simulation results are sought as scattering parameters.

The function "initialize_problem_space()" constructs the material grid of the three-dimensional problem space based on the substrate, the microstrip patches on the surface of the substrate, vias, and air surrounding the substrate. In this implementation an air buffer is created by default around the substrate. The thickness of the air buffer is at least 20 cells or 3 times the substrate thickness on five sides of the substrate except for below the ground. The material grid includes three dimensional arrays representing permittivity and electric conductivity distributions that are associated with x, y, and z components of the electric field.

Once the problem space is constructed, the "display_problem_space()" function can be used to display the geometry of the problem space including its boundaries, as well as the material grid that will be simulated in the subsequent step. Since a continuous three-dimensional space is approximated by a discrete representation, it is essential to check the geometry, as well as the material grid to make sure that everything is set up correctly in the discrete representation. For instance, **Figure 1** shows a microstrip low pass filter, which is a classic example presented in Sheen et al. [12], as displayed by the program, and **Figure 2** shows the grid including the conductivity distribution on the surface of the substrate.

As the next step, "initialize_updating_coefficients_and_fields()" function is called to create three-dimensional coefficient and field arrays. For instance, Listing 2 shows the section of the code that is used to create the general coefficient arrays based on (13). In addition to the general updating coefficients, ports and absorbing boundaries as well are set up at this stage.

Listing 2. Section of the code that creates the general coefficient arrays.

```
Cexe = (2 .* eps_r_x .* eps_0 .- dt .* sigma_e_x)
          ./(2 .* eps_r_x .* eps_0 .+ dt .* sigma_e_x);
Cexh = (2*dt*dx/(dy*dz))
          ./ (2 .* eps_r_x .* eps_0 .+ dt .* sigma_e_x);
Ceye = (2 .* eps_r_y .* eps_0 .- dt .* sigma_e_y)
          ./(2 .* eps_r_y .* eps_0 .+ dt .* sigma_e_y);
Ceyh = (2*dt*dy/(dx*dz))
          ./(2 .* eps_r_y .* eps_0 + dt .* sigma_e_y);
Ceze = (2 .* eps_r_z .* eps_0 .- dt .* sigma_e_z)
          ./ (2 .* eps_r_z .* eps_0 + dt .* sigma_e_z);
Cezh = (2*dt*dz/(dx*dy))
          ./ (2 .* eps_r_z .* eps_0 + dt .* sigma_e_z);
Chxe = dt*dx/(dy*dz*mu_0)
Chye = dt*dy/(dz*dx*mu_0)
Chze = dt*dz/(dx*dy*mu_0)
```

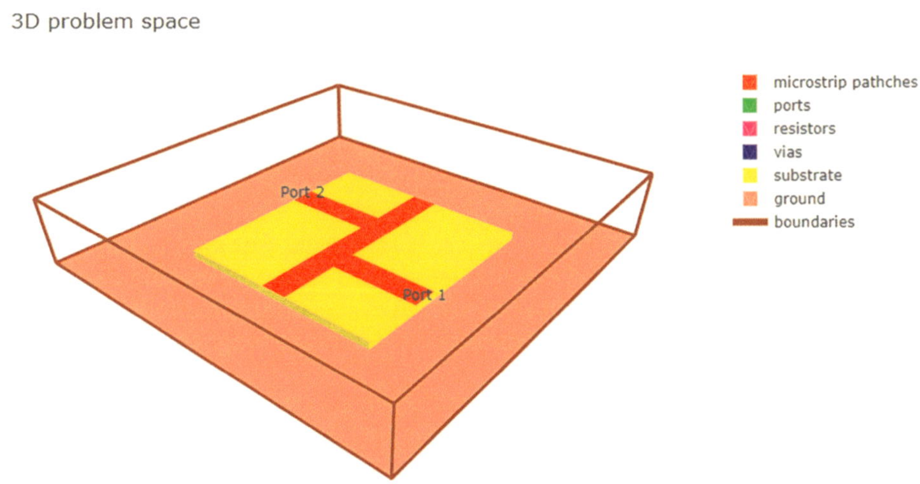

Figure 1.
The geometry of a low-pass filter presented in Sheen et al. [12].

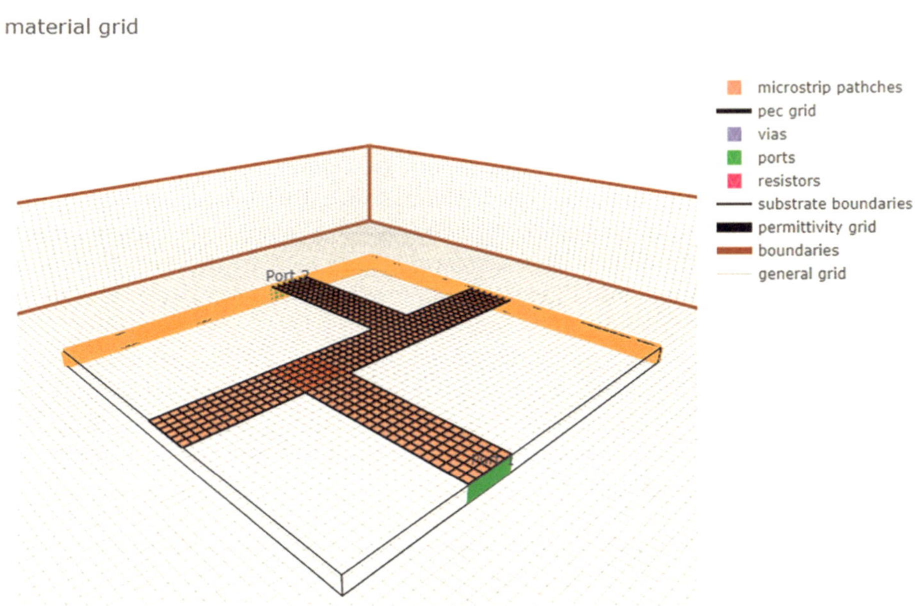

Figure 2.
Material mesh display of the circuit in ***Figure 1***.

It should be noted that all three-dimensional arrays that represent fields as well as coefficients are of data type Float32 in the presented Julia code, which is the single-precision floating-point format that occupies 32 bits in computer memory. Generally double-precision floating-point numbers are used in scientific computation due to the higher accuracy it can provide. However, our experience shows that single-precision numbers are sufficient for the most time to achieve sufficiently accurate results in FDTD simulations. In general single-precision arithmetic is faster-than double-precision arithmetic and single-precision can safely be preferred to achieve faster execution of the program in the case of FDTD.

DOI: http://dx.doi.org/10.5772/intechopen.1004059

When initializing the ports, data structures for a voltage source, a captured voltage and a captured current at the port location are created for each port. In this code the waveform to excite a voltage source is created as the derivative of Gaussian waveform by default. This waveform is a wideband waveform in its frequency spectrum while excluding very low frequencies and it can lead to faster simulation in which the transients can vanish quickly. The code can be easily modified to accommodate Gaussian, cosine modulated Gaussian, or any other custom waveform that can be used to excite a voltage source.

The next step is to run the time marching loop in "run_time_marching_loop()", part of which is shown in Listing 3. Notice that since the main goal of the FDTD analysis of a microstrip circuit is to obtain its scattering parameters, the time marching loop needs to be repeated for each port. Each time one port needs to be active as the source port while other sources are inactive. The voltage source of the active port is excited during the time marching loop, and voltages and currents are captured at all ports and stored in their respective arrays to be used during a post-processing step later. Each time the time marching iterates until transient fields are sufficiently vanished or the maximum number of time steps indicated in the project file is reached. In the given code, transient voltage that is captured at the source port is evaluated to determine if the transients vanished sufficiently; if the sum of the absolute values of voltages in the last 200 time steps is less than a value chosen as the stopping criterion then the time marching loop is ended.

Listing 3. Code showing the time marching loop executed for each active port.

```
for active_port in 1:number_of_ports
    if (!ports[active_port].is_active)
      continue
    end
    println("<Running port $active_port of $number_of_ports ports>")
    clear_fields(Ex, Ey, Ez, Hx, Hy, Hz, pml_boundary)
    vs = ports[active_port].voltage_source
    vf = v_fields[active_port]
    for time_step = 1:number_of_time_steps
        if (check_convergence(ports, active_port, time_step))
          break
        end
    update_Hx(Hx, Chxe, Ey, Ez, nx, ny, nz, nxp1, nyp1, nzp1)
    update_Hy(Hy, Chye, Ez, Ex, nx, ny, nz, nxp1, nyp1, nzp1)
    update_Hz(Hz, Chze, Ex, Ey ,nx, ny, nz, nxp1, nyp1, nzp1)
    update_Hx_PML(Hx, Chxe, Ey, Ez, nx, ny, nz,
                        nxp1, nyp1, nzp1, pml_boundary)
    update_Hy_PML(Hy, Chye, Ez, Ex, nx, ny, nz,
                        nxp1, nyp1, nzp1, pml_boundary)
    update_Hz_PML(Hz, Chze, Ex, Ey, nx, ny, nz,
                        nxp1, nyp1, nzp1, pml_boundary)
    capture_sampled_currents(ports, Hx, Hy,
                        active_port, time_step)
    update_Ex(Ex, Cexe, Cexh, Hz, Hy, nx, ny, nz,
                        nxp1, nyp1, nzp1)
    update_Ey(Ey, Ceye, Ceyh, Hx, Hz, nx, ny, nz,
                        nxp1, nyp1, nzp1)
    update_Ez(Ez, Ceze, Cezh, Hy, Hx, nx, ny, nz,
```

```
                               nxp1, nyp1, nzp1)
        update_Ex_PML(Ex, Cexe, Cexh, Hz, Hy, nx, ny, nz,
                               nxp1, nyp1, nzp1, pml_boundary)
        update_Ey_PML(Ey, Ceye, Ceyh, Hx, Hz, nx, ny, nz,
                               nxp1, nyp1, nzp1, pml_boundary)
        update_Ez_PML(Ez, Ceze, Cezh, Hy, Hx, nx, ny, nz,
                               nxp1, nyp1, nzp1, pml_boundary)
        update_voltage_sources(vs, vf, time_step)
        capture_sampled_voltages(ports, v_fields,
                               active_port, time_step)
    end
end
```

During each iteration of the time marching loop, first, magnetic fields H_x, H_y, and H_z are updated each by calling a respective function. For instance, Listing 4 shows the function "update_Hx()" to update H_x based on the Eq. (14). Notice that the coefficient "Chxe" is a scalar here.

After magnetic fields are updated in the entire three-dimensional space, magnetic fields are updated on the boundaries based on the ADE-PML boundary conditions equations. Next, values of currents are captured at all ports using the newly updated magnetic fields.

Listing 4. The function that updates H_x.

```
function update_Hx(Hx, Chxe, Ey, Ez,nx, ny, nz, nxp1, nyp1, nzp1)
    Hx[1:nxp1,1:ny,1:nz] .= (Hx[1:nxp1,1:ny,1:nz]
        .+ Chxe .* (Ey[1:nxp1,1:ny,2:nzp1] .- Ey[1:nxp1,1:ny,1:nz]
        .- Ez[1:nxp1,2:nyp1,1:nz] .+ Ez[1:nxp1,1:ny,1:nz]));
end
```

Then the iteration proceeds to compute electric fields E_x, E_y, and E_z by calling a respective function for each. For instance, Listing 5 shows the function "update_Ex()" to update E_x based on the Eq. (13). Notice that the coefficients "Cexe" and "Cexh" both are three-dimensional arrays here.

Listing 5. The function that updates E_x.

```
function update_Ex(Ex, Cexe, Cexh, Hz, Hy,nx, ny, nz, nxp1, nyp1, nzp1)
    Ex[1:nx,2:ny,2:nz] .= (Cexe[1:nx,2:ny,2:nz] .* Ex[1:nx,2:ny,2:nz]
                   .+ Cexh[1:nx,2:ny,2:nz] .* (Hz[1:nx,2:ny,2:nz]
                   .- Hz[1:nx,1:ny-1,2:nz] .- Hy[1:nx,2:ny,2:nz]
                   .+ Hy[1:nx,2:ny,1:nz-1]));
End
```

After electric fields are updated in the entire three-dimensional space, electric fields are updated on the boundaries based on the ADE-PML boundary conditions equations. Next, electric fields are updated due to voltage source excitation at the location of the active port. Then the values of voltages are captured at all ports using the newly updated electric fields.

Once the time marching loops are completed, all transient voltages and currents that are captured at all ports for all port excitations are available to perform post processing. The "calculate_scattering_parameters()" function is called to compute the scattering parameters.

As the last step of the FDTD simulation, the "display_simulation_results()" function is called to display all transient voltages and currents captured during the FDTD simulation as well as the scattering parameters computed in the post processing step. We will show such results in the next section.

4. Example circuits simulated in Julia FDTD

In this section we will present simulations of two example microstrip circuits. The first circuit is a low pass microstrip circuit that is illustrated in **Figure 2**. Listing 6 shows the XML file that is used to set up the analysis for this circuit. In this file the cell size is set as $\Delta x = 0.4064mm$, $\Delta y = 0.4233mm$, and $\Delta z = 0.265mm$. The substrate has a dielectric constant of $\varepsilon_r = 2.2$ and a thickness of $\Delta h = 0.795mm$.The maximum number of time steps to run the simulation is 4000. The size and position of the substrate and microstrip patches are listed as coordinates of rectangular shapes. Port 1 is set active while Port 2 is inactive, such as to simulate Port 1 excitation only to obtain the scattering parameters S_{11} and S_{21}. The Julia program is executed by calling “run_fdtd(“low_pass_filter.xml”)”.

For this example, the program reached the stopping criterion at about 3700 time steps. Then the simulation results are generated as plots. **Figure 3** shows the voltage waveform generated by the Port 1 voltage source as well as the transient voltages captured at Port 1 and Port 2 up to 1 nano seconds. **Figure 4** shows the scattering parameters on a frequency range 0.5 GHz to 20 GHz.

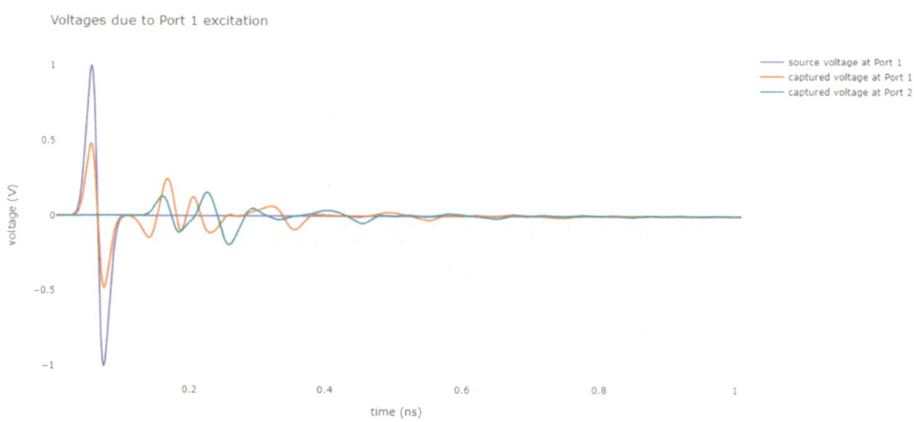

Figure 3.
Transient voltages captured due to port 1 excitation.

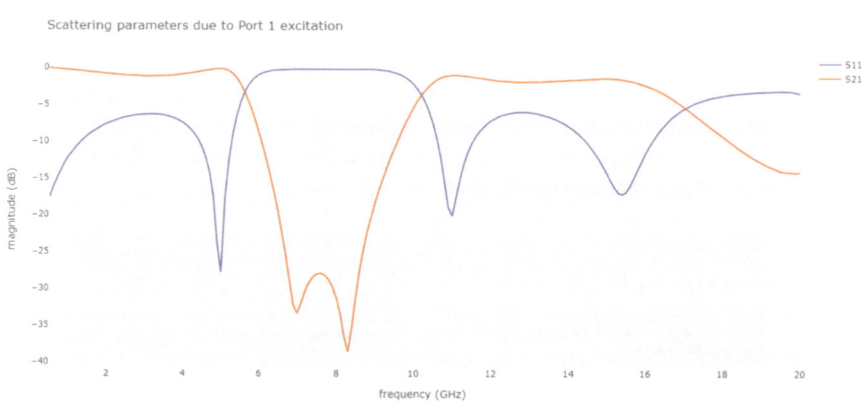

Figure 4.
Scattering parameters of the low pass filter on a range 0.5 GHz to 20 GHz.

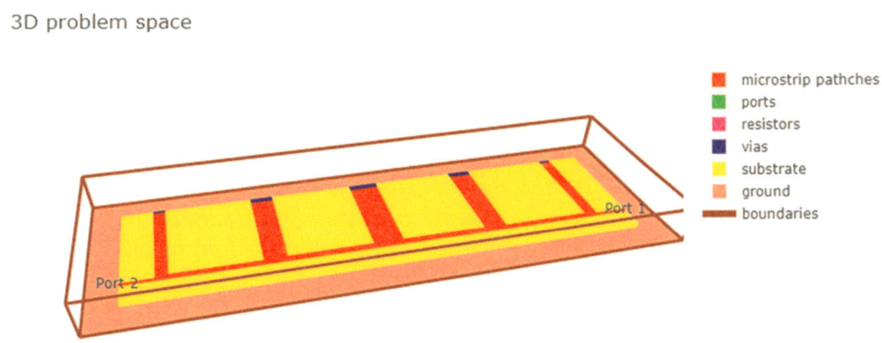

Figure 5.
The geometry of a band pass filter presented in Hong [13].

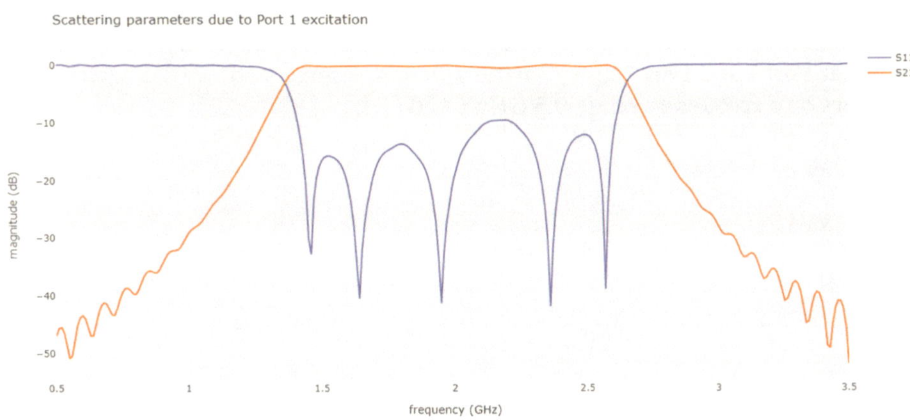

Figure 6.
Scattering parameters of the band pass filter on a range 0.5 GHz to 3.5 GHz.

The second example, shown in **Figure 5**, is a bandpass filter presented in Hong [13] (in Figure 5.27). This filter is designed based on a five-pole Chebyshev lowpass prototype with a 0.1-dB passband ripple. For the microstrip filter design, a dielectric substrate with a relative dielectric constant of $\varepsilon_r = 10.2$ and a thickness of $\Delta h = 0.635mm$ is used. Listing 7 is a section of the XML file that is used to set up simulation of this band pass filter and it shows how the geometry of the microstrip circuit is created using a polygon and multiple vias. For the FDTD simulation of this circuit a cell size of $\Delta x = 0.2mm$, $\Delta y = 0.2mm$, and $\Delta z = 0.1585mm$ is used. **Figure 6** shows the scattering parameters calculated after 40,000 time steps.

Listing 6. XML file that is used to set up simulation of a low pass filter.

```
<?xml version="1.0" encoding="UTF-8"?>
<project>
<analysis_setup>
    <dx>0.4064e-3</dx>
    <dy>0.4233e-3</dy>
    <dz>0.265e-3</dz>
    <maximum_number_of_time_steps>4000</maximum_number_of_time_
    steps>
    <courant_factor>0.9</courant_factor>
</analysis_setup>
```

```
<geometry>
    <substrate>
        <relative_permittivity>2.2 </relative_permittivity>
        <electric_conductivity>0.0 </electric_conductivity>
        <thickness>0.795e-3 </thickness>
        <min_coordinate> 0 0 </min_coordinate>
        <max_coordinate> 0.0203 0.0195 </max_coordinate>
    </substrate>
    <rectangle>
        <min_coordinate> 0 0.0085 </min_coordinate>
        <max_coordinate> 0.0203 0.011 </max_coordinate>
    </rectangle>
    <rectangle>
        <min_coordinate> 0.0056896 0.0 </min_coordinate>
        <max_coordinate> 0.008128 0.0085 </max_coordinate>
    </rectangle>
    <rectangle>
        <min_coordinate> 0.012192 0.0085 </min_coordinate>
        <max_coordinate> 0.0146304 0.0195 </max_coordinate>
    </rectangle>
</geometry>
<sources>
    <port>
          <min_coordinate> 0.012192 0.0195 </min_coordinate>
          <max_coordinate> 0.0146304 0.0195 </max_coordinate>
          <impedance> 50 </impedance>
          <is_active> 1 </is_active>
    </port>
    <port>
          <min_coordinate> 0.0056896 0.0 </min_coordinate>
          <max_coordinate> 0.008128 0.0 </max_coordinate>
          <impedance> 50 </impedance>
          <is_active> 0 </is_active>
    </port>
</sources>
<outputs>
    <frequencies>
          <start>0.5e9</start>
          <end>20e9</end>
          <step>0.1e9</step>
    </frequencies>
</outputs>
</project>
```

5. Conclusions

In this chapter we presented an implementation of the FDTD method in Julia customized to simulate microstrip circuits. The code of this implementation is accessible at [11]. The code can be downloaded and used as a base code to experiment with various features of the FDTD method and it can be optimized for more efficient execution following the best practices and performance tips for Julia.

It should be noted that the most effective method to improve the efficiency of a program is to utilize the parallel processing capabilities of the computer system on which the program is running. Julia supports four categories of concurrent and parallel programming, which are asynchronous tasks or coroutines, multi-threading, distributed computing, and GPU computing. It is highly recommended to utilize these parallel processing capabilities supported by Julia based on the available computer architecture.

Listing 7. Part of the XML file that is used to set up simulation of the band pass filter.

```
<geometry>
    <substrate>
        <relative_permittivity>10.2 </relative_permittivity>
        <electric_conductivity>0.0 </electric_conductivity>
        <thickness > 0.635e-3 </thickness>
        <min_coordinate> −0.03324 0 </min_coordinate>
        <max_coordinate>0.03324 0.018 </max_coordinate>
    </substrate>
    <polygon>
        <vertex> −0.03324 0.01441 </vertex>
        <vertex> −0.02904 0.01441 </vertex>
        <vertex> −0.02904 0 </vertex>
        <vertex> −0.02743 0 </vertex>
        <vertex> −0.02743 0.01403 </vertex>
        <vertex> −0.01594 0.01403 </vertex>
        <vertex> −0.01594 0 </vertex>
        <vertex> −0.01294 0 </vertex>
        <vertex> −0.01294 0.0139 </vertex>
        <vertex> −0.001965 0.0139 </vertex>
        <vertex> −0.001965 0 </vertex>
        <vertex> 0.001965 0 </vertex>
        <vertex> 0.001965 0.0139 </vertex>
        <vertex> 0.01294 0.0139 </vertex>
        <vertex> 0.01294 0 </vertex>
        <vertex> 0.01594 0 </vertex>
        <vertex> 0.01594 0.01403 </vertex>
        <vertex> 0.02743 0.01403 </vertex>
        <vertex> 0.02743 0 </vertex>
        <vertex> 0.02904 0 </vertex>
        <vertex> 0.02904 0.01441 </vertex>
        <vertex> 0.03324 0.01441 </vertex>
        <vertex> 0.03324 0.015 </vertex>
        <vertex> −0.03324 0.015 </vertex>
     </polygon>
<via>
     <min_coordinate> −0.02904 0 </min_coordinate>
     <max_coordinate> −0.02743 0.00036 </max_coordinate>
</via>
<via>
```

```
        <min_coordinate> −0.01594 0 </min_coordinate>
        <max_coordinate> −0.01294 0.00083 </max_coordinate>
</via>
<via>
        <min_coordinate> −0.001965 0 </min_coordinate>
        <max_coordinate>0.001965 0.00087 </max_coordinate>
</via>
<via>
        <min_coordinate>0.01294 0 </min_coordinate>
        <max_coordinate>0.01594 0.00083 </max_coordinate>
</via>
<via>
        <min_coordinate>0.02743 0 </min_coordinate>
        <max_coordinate>0.02904 0.00036 </max_coordinate>
</via>
</geometry>
```

Author details

Veysel Demir
Department of Electrical Engineering, Northern Illinois University, DeKalb, IL, The United States of America

*Address all correspondence to: vdemir@niu.edu

References

[1] Yee K. Numerical solution of initial boundary value problems involving maxwell's equations in isotropic media. IEEE Transactions on Antennas and Propagation. 1966;**14**(3):302-307

[2] Taflove A, Hagness SC. Computational Electrodynamics: The Finite-Difference Time-Domain Method. 3rd ed. Norwood, MA: Artech House; 2005

[3] Elsherbeni A, Demir V. The Finite-Difference Time-Domain Method for Electromagnetics with MATLAB® Simulations: ACES Series. 2nd ed. Edison, NJ: Scitech Publishing; 2015

[4] Sullivan DM. Electromagnetic Simulation Using the FDTD Method. Piscataway, NJ: John Wiley & Sons; 2013

[5] Gedney S. Introduction to the Finite-Difference Time-Domain (FDTD) Method for Electromagnetics. Vol. 6. 2011. pp. 227-230

[6] The Julia Programming Language. Available from: https://julialang.org/

[7] Bezanson J, Edelman A, Karpinski S, Shah VB. Julia: A fresh approach to numerical computing. SIAM Review. 2017;**59**(1):65-98. DOI: 10.1137/141000671

[8] Gedney SD, Zhao B. An auxiliary differential equation formulation for the complex-frequency shifted PML. IEEE Transactions on Antennas and Propagation. 2010;**58**(3):838-847

[9] Liao Z, Huang K, Yang BP, Yuan YF. A transmitting boundary for transient wave analyses. Science in China Series A-Mathematics, Physics, Astronomy & Technological Science. 1984;**27**: 1063-1076

[10] Mur G. Absorbing boundary conditions for the finite-difference approximation of the time-domain electromagnetic-field equations. IEEE Transactions on Electromagnetic Compatibility. 1981;**EMC-23**(4):377-378

[11] FDTD in Julia. Available from: http://www.sosine.net/

[12] Sheen DM, Ali SM, Abouzahra MD, Kong JA. Application of the three-dimensional finite-difference time-domain method to the analysis of planar microstrip circuits. IEEE Transactions on Microwave Theory and Techniques. 1990;**38**(7):849-857

[13] Hong JS. Microstrip Filters for RF/Microwave Applications. Wiley Series in Microwave and Optical Engineering. Hoboken, NJ: Wiley; 2010